Grasshoppers of Florida

INVERTEBRATES OF FLORIDA SERIES

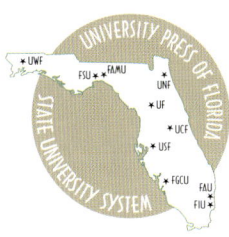

Florida A&M University, Tallahassee
Florida Atlantic University, Boca Raton
Florida Gulf Coast University, Ft. Myers
Florida International University, Miami
Florida State University, Tallahassee
University of Central Florida, Orlando
University of Florida, Gainesville
University of North Florida, Jacksonville
University of South Florida, Tampa
University of West Florida, Pensacola

INVERTEBRATES OF FLORIDA SERIES

John L. Capinera, Clay W. Scherer, and Jason M. Squitier, *Grasshoppers of Florida*

Grasshoppers of Florida

John L. Capinera, Clay W. Scherer,
and Jason M. Squitier

University Press of Florida

GAINESVILLE · TALLAHASSEE · TAMPA · BOCA RATON
PENSACOLA · ORLANDO · MIAMI · JACKSONVILLE · FT. MYERS

Copyright 2001 by John L. Capinera, Clay W. Scherer, and Jason M. Squitier
Printed in Hong Kong

06 05 04 03 02 01   6 5 4 3 2 1

ISBN 0-8130-2426-9
CIP data are available.

The University Press of Florida is the scholarly publishing agency for the State
University System of Florida, comprising Florida A&M University, Florida Atlantic
University, Florida Gulf Coast University, Florida International University, Florida
State University, University of Central Florida, University of Florida, University of
North Florida, University of South Florida, and University of West Florida.

University Press of Florida
15 Northwest 15th Street
Gainesville, FL 32611–2079
http://www.upf.com

Contents

Figures

Plates

Maps

Preface

This is the first color guide to grasshoppers of the southeastern United States. The grasshoppers rival butterflies and beetles for beauty and surpass nearly all other insect groups for their abundance and ecological importance. Grasshoppers are usually the most abundant of the large insects in all Florida ecosystems and are a critical food resource for birds and other small vertebrates. More than one-fourth of Florida's grasshoppers are unique, found nowhere else in the world. All of the acridid grasshoppers found in Florida are included in this guide. More than 100 color photographs, numerous line drawings, and nontechnical descriptions provide for easy and rapid identification. Also included is information on anatomy, life history, ecological significance, geographical distribution, and the management of grasshopper pests. This is the perfect guide for young people who are interested in collecting insects and for people of all ages interested in learning about natural history.

Acknowledgments

Preparation of this publication was supported by Florida Agricultural Experiment Station, and it is produced as publication R-05862. The line drawings were developed by Pat Hope, maps and pie charts were created by Seth Ambler, and production and editing assistance was provided by Jane Medley, all of the Department of Entomology and Nematology, University of Florida.

I

Introduction

An Overview of Grasshoppers and Their Biology

 The purpose of this guide to the grasshoppers (family Acrididae) found in Florida is to describe and portray the species in a manner that allows ready identification by the reader. It is designed for anyone interested in natural history, not just scientists and future scientists. In striving to make it user friendly, we have attempted to minimize scientific terminology, and we provide definitions and drawings to increase comprehension and ease of use when scientific terminology does occur. Where possible, we have used both the common, or English, name and the scientific name for both grasshopper species and subfamilies. Most of the common names originate with Blatchley (1920) or Helfer (1972). There is nearly universal agreement on the scientific names, but not on the common names. Therefore, it is really better to use the seemingly unpronounceable scientific names, because at least there is some consistency and they are recognized everywhere in the world. Scientific names consist of a genus, a species, and an author. A genus consists of a group of closely related species—but not so closely related that interbreeding is either possible or normally occurring between the different species. The genus name is always capitalized. The term *species* is reserved for populations of organisms that are so closely related they can interbreed. The species name follows the genus name and is not capitalized. Someone has given a scientific description of each species. This describer is called the author, and his or her name follows the genus and species designation. Sometimes the author's name is placed in parentheses, indicating that there has been a change in the name since its original description. If the author's name does not appear in parentheses, this indicates that the name is exactly as it appeared when the species was first described.

The body length measurements provided in the text are total maximum body length, including wings but excluding antennae and legs. Therefore, for longwinged species the measurements are taken from the front of the head to the tip of the folded wings. For shortwinged species, of course, the measurements are taken from the front of the head to the tip of the abdomen, or last body segment. All measurements are given in millimeters (mm); if you would like to use inches, just remember that there are 25 mm to the inch.

Grasshoppers are among the largest and most plentiful of insects, yet they are poorly known. With this guide you can identify all seventy of the grasshopper species that inhabit Florida, and many species also found in nearby states. You also can learn about their natural history, their distribution within the state, and how to collect and preserve them. It is hard to imagine another group of insects so accessible and identifiable, yet providing so exciting a challenge for collectors. To collect examples of all of Florida's grasshoppers, one must be willing to cross the state from the Panhandle to the Lake Okeechobee area and endure the rigors of Florida's hot, humid, sandy, marshy, and densely forested habitats. Because many of Florida's grasshoppers are poorly known, you can make a real contribution to the scientific study of insects by recording where these species are found, what they feed upon, and the time of the year they are abundant.

A central repository for insects and insect biology is the collection of the state insect museum, which consists of more than seven million insects:

Florida State Collection of Arthropods
Florida Department of Agriculture and Consumer Services
Division of Plant Industry
1911 SW 34th Street
Gainesville, Florida 32608

What Is a Grasshopper?

There are several closely related groups of insects in the order Orthoptera that are sometimes called grasshoppers. However, most entomologists and nonentomologists consider only insects in the family Acrididae, a smaller division of the order Orthopera, to mean "grasshopper." Indeed, we consider only members of this family in our treatment of Florida grasshoppers. Acrididae are sometimes referred to as the shorthorn grasshoppers, a reference to their relatively short antennae. However, other families that are sometimes thought of as grasshoppers include the family Tetrigidae, or pygmy grasshoppers; the family Eumastacidae, or monkey grasshoppers; the family Tanoceridae, or desert longhorn grasshoppers; the family Tettigoniidae, or longhorn grasshoppers or katydids; the family Gryllacrididae, or wingless longhorn grasshoppers; and perhaps other small families. Other than the acridids, only the tettigoniids are known to most people, and they are usually known as katydids or coneheads. Of the grasshoppers and their close relatives, only Acrididae and Tettigoniidae are common in Florida.

Acridid grasshoppers usually are large insects. Their antennae are relatively short, often less than half the length of the body. Acridids may be winged or wingless, but if winged they have four wings: a pair of front wings, customarily called forewings, and a pair of hind wings. Wing size varies considerably. Shortwinged forms are flightless, whereas longwinged forms are sometimes strong fliers. The forewings are somewhat thickened and pigmented, or colored. They are called tegmina. The hind wings are not thickened, and may range from unpigmented to brightly colored. The hind wings are often large and fan-shaped, and fold up under the forewings when the insect is not in flight. Grasshoppers tend to have long legs. The hind legs are especially elongate and enlarged to facilitate leaping, as well as armed with spines for defense.

Grasshoppers produce sound by rubbing one part of the body against another, though the parts involved may vary from one species to the next. Their hearing is often aided by the presence of tympana, auditory organs on the sides of the first abdominal segment, just behind the hind legs. Singing is performed principally by males as part of their courtship ritual.

The Parts of a Grasshopper

The morphology, or body details, of grasshoppers can seem complex, especially because of the foreign-sounding names affixed to many structures. However, one need examine the color and shape of only a few structures to accurately identify grasshoppers. It is not much more difficult than identifying birds, and often does not require great magnification. Grasshoppers can be identified with no more than a 10 × hand lens, which is readily available from any hobby shop. We have attempted to simplify the identification of grasshoppers by minimizing the use of terminology and by providing a glossary, in the back of the book, and drawings of the body parts.

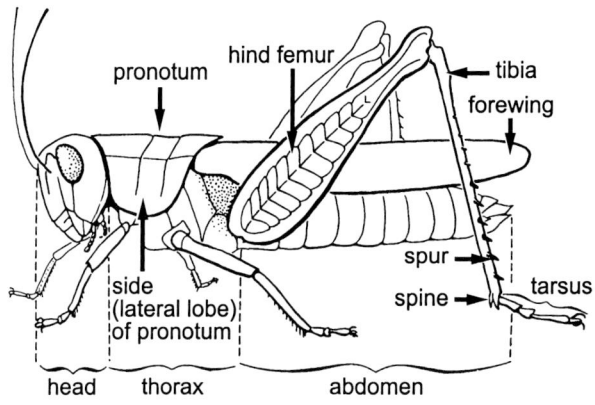

Figure 1. Parts of adult grasshopper, side view.

The grasshopper body is divided into three basic components: the head, which bears the sensory structures such as eyes, antennae, and mouthparts; the thorax, which bears the structures associated with movement, namely the legs and wings; and the abdomen, which bears the digestive and reproductive structures. A few elements associated with each of these body segments provide the key structures for grasshopper identification.

On the grasshopper head, the principal structures used for identification are the antennae, sensory structures attached to the front of the head between the eyes. Most grasshopper antennae consist of a string of small, barrel-like segments, although the individual segments are sometimes quite elongate. Some species, however, have flattened antennal segments, and many have flattened segments that are larger toward the base of the antenna and smaller toward the tip. Such antennae are said to be sword-shaped. A feature that is important in a few species is the shape of the vertical, flattened, elevated structure on the front of the head, called the frontal ridge.

The thorax consists of three segments, although this is not readily apparent because the first of the three segments, the prothorax, is enlarged and covers the other segments when viewed from above. The prothorax often bears longitudinal ridges, sometime called carinae. The shape of the medial or central ridge, or the paired side ridges, is sometimes a key character in identification. The ridges may be cut by crevices, called sulci. One pair of legs is attached to each of the three segments of the thorax, with the third pair, or hind legs, enlarged. The important hind leg segments, from the perspective of identification, are the large, thickened femur (plural, femora) and the long, thin tibia (plural, tibiae). Attached to the second and third segments of the thorax are the wings (if they are present). The forewings, also called tegmina, attach to the middle, or second, thoracic segment, and tend to be narrow, thickened and pigmented. The forewings overlay the hind wings, which are attached to the posterior, or third, thoracic segment. The hind wings usually are broad, thin, and transparent. The hind wings provide most of the lift used for flight, and remain folded and unseen until the insect flies. Wing length varies within some species,

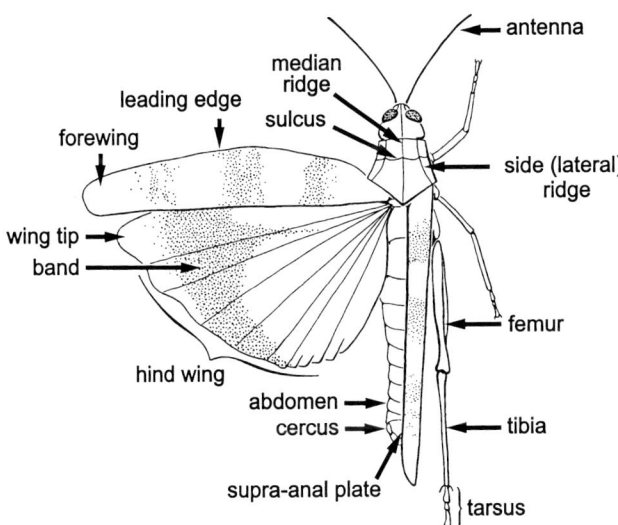

Figure 2. Parts of adult grasshopper, top view.

but is commonly used to distinguish among the various species. Species that inhabit open environments where flight is easy, such as pastures and marshes, usually are longwinged (macropterous). In contrast, species that inhabit dense vegetation and undergrowth, where flight is perhaps more difficult, are more likely to be shortwinged (brachypterous).

The abdomen is the largest and hindmost component of the grasshopper body. It bears the reproductive structures at the very end. In some groups, small paired appendages called cerci are species-specific in shape among males. Another structure that aids identification is the furcula, a forked organ in which only the two tips of the fork are visible, making it appear that there are two structures rather than one. In males, the furcula rests on a broad, flat plate at the top of the abdomen near its tip, called a supra-anal plate. The tip of the abdomen in males is called the subgenital plate. It is mostly on the underneath side of the abdomen and at its tip, topped by the supra-anal plate. The shapes of both the supra-anal plate and subgenital plate sometimes have value in making identification. Beneath the supra-anal plate of males is the aedeagus, or penis. We do not

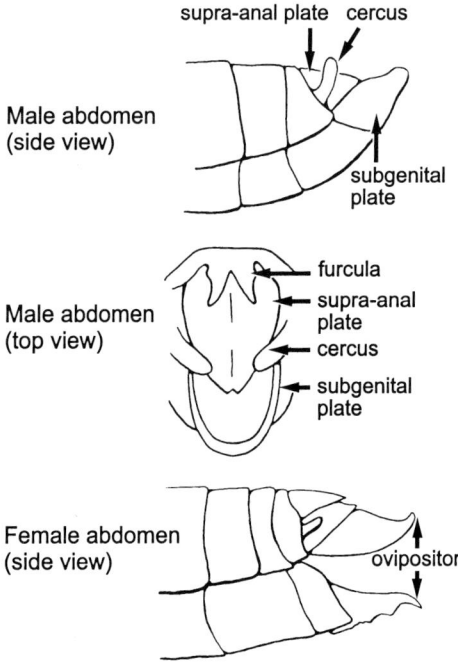

Figure 3. Tip of abdomen in adult male and female grasshoppers.

describe this structure because it is internal, and examination requires difficult manipulation of the specimen and significant magnification. However, the form of the aedeagus may be critical in sexual compatibility, and its shape is an excellent indicator of identity in some grasshoppers.

In females, the tip of the abdomen is dominated by the ovipositor, a structure that is inserted into the soil and used to dig a hole to prepare for egg laying. It consists of curved, pointed structures that open upwards and downwards. The ovipositor has great value in distinguishing the sex, but very limited value in species identification.

The Life History of Grasshoppers

The life history of grasshoppers is relatively simple, although it varies somewhat among different species. The principal stages are the egg, nymph, and adult. Grasshoppers undergo gradual metamorphosis (the nymph gradually changes to the adult form), in contrast to insects that undergo complete metamorphosis (there is a pupal stage between the immature and adult stages).

Adult females produce eggs, which are deposited in clusters, usually in the soil. Clusters of eggs are usually held together by a frothy secretion that, when dry, forms a rigid covering over the eggs. The eggs and frothy secretion are collectively known as an egg pod. A pod may contain 4 to more than 100 eggs, depending on the grasshopper species. Grasshoppers typically pass winter in the egg stage. In Florida, however, many species survive the winter months in the nymphal and adult forms. This is especially true in the southern half of the state.

When a grasshopper egg hatches, the young grasshopper digs its way through the soil to the surface and molts into an active form capable of walking, hopping, and eating. The active stage between hatching and adulthood is called a nymph, with stages between molts called instars. This first active form, known as the first instar, is followed by additional molts until it has experienced (usually) five or six instars, and is ready for its final molt to the adult form. The principal reason that insects molt, or shed their old body covering, is because the covering is not elastic and inhibits growth. Thus, each time the grasshopper nymph molts it produces a larger body covering, and the nymph gets larger and larger. As grasshopper nymphs grow the wings begin to develop, but they are not fully formed until the adult stage. Sexual structures, such as the ovipositor in females, also develop as the grasshoppers grow. These, too, are not fully formed until the adult stage.

The nymphal instars can be distinguished by the shape of the wing pads, or partly developed wings. The wing pads initially are very short and broadly rounded, but become slightly elongated in instar 2. Instars 3 and 4 have more elongate, downward pointing wing pad tips, and display some weak wing veins. In instars 5 and 6 the wing pads are inverted; they point upward or back instead of downward, and only one pair of wings is visible. In grasshoppers with only five instars, the pattern is much the same, but the fourth nymphal stage is absent from the developmental sequence.

Adult grasshoppers can reproduce, and have fully formed sexual organs, some of which are visible externally. They also have fully formed wings. Many species are macropterous, or longwinged, which means that the wings extend nearly to the tip of the abdomen or beyond. Many of Florida's grasshoppers, however, are brachypterous, or shortwinged, in the adult stage. Such wings are typically oval and extend only about one-third the length of the abdomen. A few species are wingless, or nearly so, in the adult stage. It can be difficult to distinguish between immatures and adults when adults can be longwinged, shortwinged, and wingless! However, if you look for obvious, fully formed genitalia, and wings extending at least one-third the length of the abdomen, you will identify most adults accu-

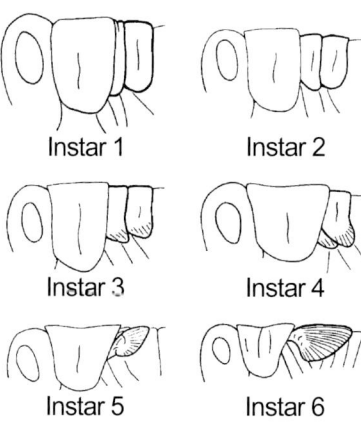

Figure 4. Front region of grasshopper nymphs showing wing characters that are used to distinguish instars in "typical" species.

rately. Only in a few species are adults likely to look like immatures, and they are easily recognized from pictures.

In the "typical" grasshopper life cycle, eggs hatch in the spring, nymphs develop through the summer, adults mate and produce eggs in the late summer and autumn, and the winter is passed in the egg stage. How many grasshoppers conform to this scenario? In northern Florida, most species seem to conform. In southern Florida it is not uncommon to have nymphs and adults present nearly year round. Unfortunately, grasshoppers have been poorly studied in Florida, so in most cases we do not know much about seasonal development. In the case of *Schistocerca americana*, the American grasshopper, we know that there are two generations annually, with adults overwintering. Clearly this is not a "typical" grasshopper, but how many other species have interesting and unusual biology? Your field observations may help us learn the biologies of other species.

The Ecological Significance of Grasshoppers

Grasshoppers often appear to be the most abundant aboveground insects. This is especially true in open, sunny, dry habitats such as prairies and pastures, but it also sometimes applies to open woods, salt marshes, and disturbed areas such as crop fields. Grasshoppers exert ecological impact and may be the dominant herbivores, or plant-feeders, in some communities.

Plant feeding by grasshoppers can deplete plant biomass and damage crops. It can shift plant-community structure because of grasshoppers' preference for one kind of plant over another. In extreme cases, herbivory can cause ecosystem damage, either directly, from disruption of habitat by loss of vegetation, or indirectly, through increased erosion caused by reduced vegetation. Such habitat damage is rare, however, especially in areas with high rainfall, such as the southeastern states.

Grasshoppers are significant due to their numbers and to their role in nutrient cycling. Grasshoppers consume large amounts, often eating their body weight in plant tissue daily. The consumption of plant tissue affects the relative abundance of different plant species in an area, due to selective feeding behavior by grasshoppers. Grasshoppers also hasten the degradation of cellulose and other materials by breaking up the plants into smaller pieces that can be attacked by soil flora and fauna. Grasshopper fecal material, in particular, is easily degraded, resulting in increased solubility of chemical nutrients essential for plant growth. Degradation of fecal material and clipped foliage causes rapid release of nutrients into the soil, favoring new plant growth. Without plant feeders such as grasshoppers, much of the nutrients in an area would be bound up in dead plant tissue, insoluble, and unavailable for plant uptake. Nutrient cycling is especially important in the warm, sandy soils commonly found in Florida, because they are inherently nutrient-poor.

Grasshoppers are also ecologically significant because they convert plant tissue into their own body tissue—large "bite-sized" units of animal material—and thus themselves serve as food for vertebrate animals. Animal tissue is much more nutritious than plant material, especially for young and rapidly growing animals, which need the high levels of protein and lipids found in grasshoppers. Grasshoppers are large enough, and abundant enough, that they attract the attention of large numbers of vertebrate animals such as reptiles, birds, skunks, raccoons, foxes, and mice, which regularly consume them. For insect-feeding birds, such as meadowlarks and cattle egrets, grasshoppers are often the principal element of the diet, and these birds' survival and reproductive efficiency may be directly related to abundance of grasshoppers. Other species, such as kestrels and bluebirds, feed extensively on grasshoppers but readily switch to other insects if grasshoppers are in short supply. Some birds, such as sparrows, feed heavily on vegetable matter, principally seed, but feed their young on insects almost exclusively. The accompanying data on items consumed by some common birds (cattle egret, chipping sparrow, and shrike) clearly show how important grasshoppers are to birds. These data, which are based on examination of stomach contents, also show how importance of grasshoppers as a dietary component varies among species of birds.

Grasshoppers are large enough and abundant enough that hunting exclusively for grasshoppers is an energetically efficient activity for many bird species. Grasshoppers are 50–75% crude protein. Without grasshoppers to consume, many vertebrate animals would suffer from lack of a suitable source of animal protein. In some parts of the world, such as sub-Saharan Africa, grasshoppers are also a component of the human diet.

Many of Florida's grasshoppers contribute substantially to biodiversity. In fact, 18 of the grasshopper species found in Florida are precinctive

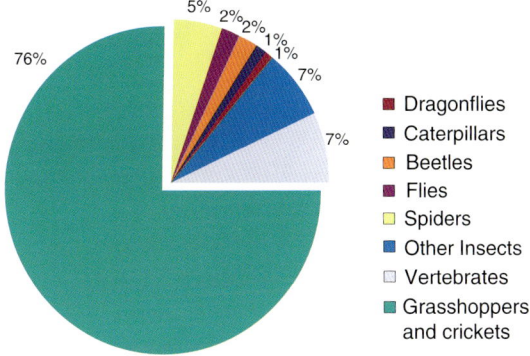

Figure 5. Cattle egret diet, showing major constituents and the average proportion of each.

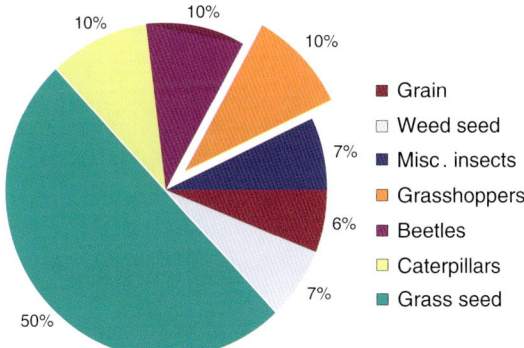

Figure 6. Chipping sparrow diet, showing major constituents and the average proportion of each.

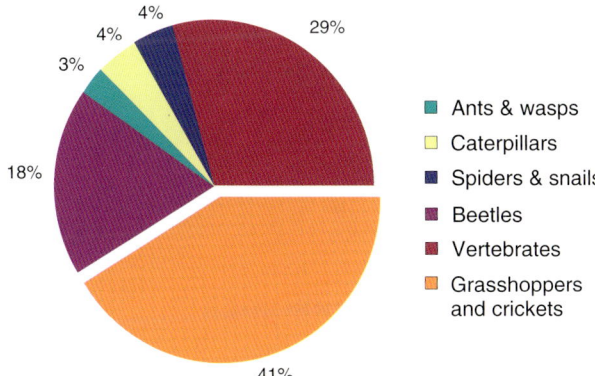

Figure 7. Shrike diet, showing major constituents and the average proportion of each.

(sometimes called endemic or indigenous)—found nowhere else in the world. The species unique to Florida are:

Eotettix palustris (Morse)—Swamp eastern grasshopper
Eotettix signatus Scudder—Handsome Florida grasshopper
Eritettix obscurus (Scudder)—Obscure slantfaced grasshopper
Hesperotettix osceola Hebard—Osceola's grasshopper
Melanoplus adelogyrus Hubbell—St. Johns spurthroat grasshopper
Melanoplus apalachicolae Hubbell—Apalachicola spurthroat grasshopper
Melanoplus davisi (Hebard)—Davis' oak grasshopper
Melanoplus forcipatus Hubbell—Toothcercus spurthroat grasshopper
Melanoplus gurneyi Strohecker—Gurney's spurthroat grasshopper
Melanoplus indicifer Hubbell—Spinecercus spurthroat grasshopper
Melanoplus nanciae Deyrup—Ocala clawcercus grasshopper
Melanoplus ordwayae Deyrup—Trail ridge scrub grasshopper
Melanoplus puer (Scudder)—Florida least spurthroat grasshopper
Melanoplus pygmaeus Davis—Pygmy spurthroat grasshopper
Melanoplus symmetricus Morse—Symmetrical spurthroat grasshopper
Melanoplus tequestae Hubbell—Tequesta spurthroat grasshopper
Melanoplus withlacoocheensis Squitier and Deyrup—Withlacoochee grasshopper
Schistocerca ceratiola Hubbell and Walker—Rosemary grasshopper

Thus, about 25% of Florida's grasshoppers are unique to Florida. They are a rich biotic resource that is not readily available to other peoples of the world. Florida's citizens and government agencies concerned about biotic diversity should recognize the ecological significance of this poorly known group of animals. Unlike many other groups of insects, currently there are no exotic or introduced species among Florida's grasshoppers. Many species of grasshoppers found in Florida have a fairly wide geographic range, often the southeastern states or the entire area east of the Rocky Mountains.

In addition to the species that are truly unique to Florida, there are several other species that have been found in nearby states, usually southern Georgia, but whose range is mostly restricted to Florida. This is the case for at least six species, and possibly others:

Gymnoscirtetes morsei Hebard—Morse's wingless grasshopper
Gymnoscirtetes pusillus Scudder—Little wingless grasshopper
Hesperotettix floridensis Morse—Florida purplestriped grasshopper

Melanoplus furcatus Scudder—Larger forktailed grasshopper
Melanoplus rotundipennis (Scudder)—Roundwinged spurthroat grass-
  hopper
Spharagemon crepitans (Saussure)—Crepitating grasshopper

It is interesting that more than 90% of the species restricted to Florida
(17 of 18 species) or nearly restricted to Florida (5 of 6 species) are in the
subfamily Cyrtacanthacridinae, and that most of these are in the genus
Melanoplus. This genus seems to be prone to formation into species, and
throughout North America there are species of *Melanoplus* with limited
geographic distribution. The southeastern and far western states seem to
support a disproportionate number of such species, however, probably
due to the type of habitat. The heavily forested southeastern states tend to
have an abundance of shortwinged species among the precinctive species.
Of the 18 precinctive species, only one or two are longwinged and capable
of flight. Prevailing thought on the evolution of wing length suggests that
grasshoppers occupying open habitats such as grasslands (campestral species) make effective use of long wings, whereas those dwelling in woods and
thick underbrush (sylvan species) obtain little advantage from flight, and
tend to have short wings. Thus, forested habitat leads to reduced flight,
which leads to reduced genetic interchange, resulting in enhanced formation of new species.

The trend toward frequent speciation, or evolution of new species,
among shortwinged grasshoppers in Florida may be reinforced by the
occurrence of "biological islands." Florida has some isolated plant communities associated with sandy ridges remaining from earlier times when
the level of the ocean was higher. The speciation associated with these
island remnants rivals the extraordinary pattern of evolution found among
finches in the Galapagos Islands, a phenomenon made famous by the
pioneering naturalist and evolutionary biologist Charles Darwin. This
interesting pattern has not been well documented or popularized in Florida,
and is still awaiting the attention of ambitious local biologists. See the
section on "What Is a Species?" for further discussion of this topic.

Grasshopper Habitats

There are few habitats in Florida that do not support grasshopper populations. The general rule is that if low-growing plants are present, grass-

hoppers can be found feeding on them. We are aware of only two habitat types where it is difficult to find grasshoppers. The first is mature, dense forest, especially pine plantations and tropical hardwood forests. In such habitat virtually no light reaches the soil, and there is no vegetation at ground level on which grasshoppers can feed. This does not mean that grasshoppers are totally absent, because small breaks in the canopy are inevitable, allowing sunlight to reach the ground, and grasshoppers locate and inhabit these areas where low-growing vegetation exists. Also, some grasshopper species feed on pine or broadleaf tree foliage, but these are not numerous and are difficult to observe.

The other habitat where grasshoppers are rare is swamps, both mangrove swamps and freshwater swamps. Although strong fliers such as the *Schistocerca* species may sometimes be observed within mangroves, they are infrequent. Mangroves inhabit coastal areas, where their roots usually are immersed in water, usually salt water. Freshwater swamps are often associated with rivers, which provide seasonal flooding. Flooding makes swamps inhospitable to grasshoppers, because nearly all grasshoppers deposit eggs in soil.

The ecosystems of Florida are extremely diverse, and have been classified in several ways depending on various authors' assessments of the significance of soil, plant communities, water relations, topography, etc. Some ecosystems represent very small areas of Florida, or differ only slightly from related ecosystems. We present here a summary of the principal natural terrestrial ecosystems in Florida, followed by some widespread artificial ecosystems, and a discussion of their suitability for grasshoppers:

1. Upland Ecosystems

Pine Flatwoods. This is the most extensive of the terrestrial ecosystems in Florida. It is characterized by low, flat topography and poorly drained, acidic, sandy soil. Pine flatwoods are open forests of longleaf or slash pine, with an understory of saw palmetto, gallberry, runner oak, wire grass, and other forbs and grasses. The pine flatwoods ecosystem is naturally maintained by frequent fire. This habitat is moderately suitable for grasshoppers, depending on the density of understory vegetation, especially palmetto, which may crowd out plants more suitable for grasshopper food.

Plate 1. Pine flatwoods habitat.

Dry Prairie. Dry prairies are similar to pine flatwoods, except that the pine trees are absent. This ecosystem historically was maintained by fire. The plants are denser now than in presettlement times, probably because uncontrolled fires are less frequent. This habitat ranges from moderately suitable to very suitable for grasshoppers. Suitability is directly related to the abundance of grass and forbs, because saw palmetto and the other shrubs found in this ecosystem are not very suitable food sources for most grasshoppers.

Scrub. Scrub is an extremely dry community with dense vegetation. Typically it consists of scrub oaks and rosemary, with or without pine trees, growing on well-drained, sandy, infertile soil. Plant density can be quite high. Dominant species often include sand pine, sand live oak, myrtle oak, Chapman's oak, saw and scrub palmetto, and Florida rosemary. The ground cover consists of litter and lichens on bare soil, and the condition

Plate 2. Scrub habitat.

is maintained by infrequent fire. Scrub areas are fairly limited in scope. The largest areas are in the vicinity of Ocala National Forest and in the area just north of Lake Okeechobee, but small areas exist throughout the state, especially along the central east coast and the coastal area of the Panhandle region. Grasshopper density tends to be low in scrub areas, but some of the more interesting and rare species occur in this habitat.

High Pine. This is an open ecosystem, with longleaf pines providing a fairly open canopy and the soil covered with grasses, commonly wire grass. High pine areas have well-drained soil and may support oaks, especially turkey oak, at various densities. High pine areas are common throughout northern and central Florida south to about Sebring, near Lake Okeechobee. This ecosystem is maintained by frequent but low-intensity fire. High pine habitats support moderate to high densities of grasshoppers because wire grass and other abundant grasses and forbs are good hosts for many grasshopper species. High pine is also called sandhill habitat, which is equally descriptive. Associated with high pine, and to a lesser degree scrub habitat, are relatively small areas with poor drainage that are called cutthroat seeps. Seeps are characterized by open slash pine forest with a thick ground cover of grass, including cutthroat grass.

Plate 3. High pine (sandhill) habitat.

Plate 4. High pine (cutthroat seep) habitat.

Plate 5. Temperate hardwood habitat.

Temperate Hardwood. In Florida, hardwood forests are called "hammocks." Temperate hardwood forests are located principally along the northern border of Florida in the Panhandle region, and the northwestern region of peninsular Florida south to about Tampa. This ecosystem is highly diverse, with numerous species of trees being predominant at any location, and the dominant species varying among locations. Oaks, southern magnolia, black cherry, dogwood, black gum, sweetgum, pignut hickory, and mockernut hickory usually dominate, but pines also may be present. Mature hardwood forest shades out grasses and most other suitable grasshopper food, so densities of these insects tend to be low.

South Florida Rockland. The elevated areas of southern Florida are usually associated with outcroppings of limestone. The soils are organic and only slightly acid. This represents a small area of Florida. Most of this ecosystem between Miami and Homestead, and in the Florida Keys, has been converted to residential or agricultural uses. There are two rather distinct habitats found within this ecosystem: tropical broadleaf hammock and pineland. Tropical broadleaf hammock is dominated by evergreen or semi-evergreen tropical broadleaf trees such as gumbo limbo, pigeon plum, wild tamarind, strangler fig, and live oak. Vines, small trees, and shrubs are common, but due to the dense foliage there are few low-growing plants and few grasshoppers. The other habitat is pineland, or Everglades flatwoods, which has an open overstory dominated by slash pine, a few shrubs and broadleaf herbs, and an abundance of grasses such as bluestem, muhly, and windmill grass. Although low in elevation, these sites remain wet for only brief periods. Pineland is maintained by frequent fire. The pineland habitat, in contrast to the tropical broadleaf habitat, supports an abundance of grasshoppers.

2. Freshwater Wetlands

Swamps. Swamps are characterized by standing water and saturated soil for at least part of the year, and usually are found along rivers. In Florida, swamps may experience fire. Swamps are quite diverse, and difficult to

Plate 6. Swamp habitat.

Plate 7. Marsh habitat.

Plate 8. Marsh habitat.

characterize. They are found throughout the state, except for the southeastern region. The most common tree associated with swamps in Florida is cypress, but palms, black gum, red maple, oaks, willow, and punktree are common elements. Various shrubs, vines, epiphytes, and even insectivorous plants can be common in this habitat. Aquatic insects, but not grasshoppers, are common in this wet environment.

Marshes. Marshes are similar to swamps in that they tend to have standing water for part of the year, and to be dominated by emergent vegetation. However, marshes tend to lack trees and often form in low areas in the absence of rivers. Marshes are most common in the central part of the state, from Orlando south to Homestead. They are most expansive south of Lake Okeechobee, an area known as the Everglades.

The vegetative composition of marshes varies considerably. Highland marshes tend to have greater diversity, possessing maidencane, cordgrass, beakrush, saw grasses, water lily, arrowhead, and many others. At the other end of the spectrum is the Everglades, where a number of species occur, but saw grasses dominate. Marsh plants, particularly saw grass, are well adapted to fire. Certain species of grasshoppers feed on marsh plants, including some that seem to be well adapted for feeding on emergent grasses.

3. Coastal Ecosystems

Salt Marsh. Salt marshes occur along both the Atlantic and Gulf coasts, but only in limited areas. Along the Atlantic coast they occur south to about Merritt Island, or the latitude of Orlando. Along the west coast

Plate 9. Salt marsh habitat.

Plate 10. Mangrove swamp habitat.

they occur north of Tampa, especially in the Big Bend region, and intermittently west to Pensacola. East coast marshes are dominated by smooth cordgrass, whereas Gulf coast marshes contain mostly black needlerush. Salt marshes provide good habitat for a few species of grasshoppers, which can be moderately abundant.

Mangrove Swamp. Mangrove swamps tend to replace salt marshes in the southern portion of the state, although mangroves may occur along most of Florida's coast. Mangrove swamps are especially extensive in southwest Florida. The dominant plants in this habitat are red, black, and white mangroves. Regular immersion of mangrove swamps by salt water and lack of plants favored for feeding make this environment relatively unsuitable for grasshoppers.

Coastal Uplands. Along most of Florida's seacoast can be found dynamic, shifting beach dune areas. Inland from these are stabilized berms or coastal strand. The more unstable areas are characterized by grasses such as sea oats, beach cordgrass, sand spur, panic grass, and seashore paspalum, and broadleaf plants such as beach morning glory, railroad vine, beach sunflower, and sea rocket. Slightly inland, shrubby plants such as sea grape, cabbage palm, evening primrose, prickly pear, Spanish bayonet, and wax myrtle also occur. Farther inland, other salt-tolerant shrubs dominate, including saw palmetto, cabbage palm, yaupon, sea grape, lantana, and woody goldenrod. Plant density increases from the beach area inland. Grass-dominated areas often support many different

Plate 11. Coastal upland (beach dune) habitat.

Plate 12. Coastal upland (coastal strand) habitat.

Plate 13. Tomato crop habitat.

kinds but moderate numbers of grasshoppers. As is the case with most plant communities, however, as they get more crowded, low-growing plants are shaded out and grasshopper abundance decreases.

4. Agricultural or Developed Ecosystems

Crops and Pastures. Crop and pasture habitat consists of a monoculture, or growth of only a single species, of introduced plants. However, both crops and pastures often have grass and broadleaf weeds interspersed among the cultivated plants. There also may be a considerable amount of bare soil present, particularly in row crops. Crop and pasture environments range from temporary to long duration. In Florida, it is often possible to grow two or even three crops annually on the same plot of ground, but perennial crops such as citrus and pasture may persist relatively unchanged for a decade or more. It is difficult to generalize about the suitability of crops and pastures for grasshoppers, but often crops are not particularly good habitats. Pastures are usually slightly better, sometimes good sites, although mostly for grass-feeding species. A key element determining the abundance of grasshoppers in crops, and to a lesser extent

Plate 14. Citrus grove habitat.

Plate 15. Young citrus grove habitat.

in pastures, is the abundance of weeds among the cultivated plants and adjacent areas. Weedy crops and pastures, and crops adjacent to abandoned cropland or weedy fence rows and irrigation ditches, tend to contain more grasshoppers.

Urbanized Land. Urbanization, whether due to construction of roads, recreation fields, or homes, has an effect on grasshoppers similar to the aforementioned crop and pasture habitats. Grasshopper species associated

Plate 16. Pasture habitat.

Plate 17. Roadside habitat.

with lawns, recreational fields, and roadsides often are the same grass-feeders associated with grass pastures. If broadleaf weeds are allowed to invade the grass habitats, or the grass remains unmowed all season, then additional species and higher numbers of grasshoppers develop.

Fallow and Disturbed Land. Crops in Florida often are subject to soil-borne disease and nematode problems. A common practice to alleviate these problems is to move the crop to land that has not supported crops for a period of years. Thus, there often is considerable acreage that supported crops in previous years but is currently idle. This is called fallow land. The fallow land may be plowed periodically to suppress weeds, but often is left unattended for long periods, allowing annual grasses and weeds, and eventually perennial plants, to grow. A nearly identical process occurs when land is temporarily disturbed, as when trees are harvested, new pine plantations are established, or soil is removed for construction. This invasion of annual plants is the first stage of ecological succession. Such weedy land may develop moderate to high grasshopper population densities. Certain grasshopper species tend to do best in disturbed habitats and use the weed flora associated with these sites, and many can disperse into nearby crops and cause injury.

Plate 18. Unused crop (fallow land) habitat.

Plate 19. Old field habitat.

Old Fields. Within a few years of cultivation or disturbance, if the land is not again disturbed, the early successional annual and perennial plants will be invaded by shrubs and trees. This complex mixture of vegetation, which can be moderately dense but not excessively tall or shaded, is called an old field habitat, and is a later stage of succession. This habitat often supports the most diverse assemblage of grasshopper species, and the highest density of grasshoppers, of any habitat. As the shrubs and trees mature, however, the habitat becomes less suitable for grasshoppers.

Grasshoppers as Pests

Grasshoppers consume considerable amounts of foliage during their nymphal development, and also as adults. When they are especially abundant they can damage economically important plants. In Florida, *Romalea microptera* and *Schistocerca americana* are the most serious grasshopper pests, occasionally damaging vegetables, citrus, and ornamental plants. Several other species cause minor damage to forage, field, and ornamental crops. Grasshoppers in Florida rarely gain the notoriety that they have attained in states west of the Mississippi River, where the arid climate allows more frequent development of grasshopper plagues. However, even in Florida, grasshoppers sometimes reach alarmingly high and damaging densities. Grasshopper management in Florida depends on both cultural and chemical techniques.

The Origin of Grasshopper Plagues

Abnormally high densities of grasshoppers are called outbreaks by entomologists, and plagues by the general public. Irrespective of the terminology applied, the phenomenon occurs throughout the world, and its origin is invariably related to food and weather.

Grasshoppers require warm, sunny conditions for optimal growth and reproduction. Warmth alone seems to be inadequate. Even during Florida's hot summer weather, grasshopper activity diminishes during cloudy weather. Thus, drought stimulates grasshopper population increase, apparently because there is less rainfall and cloudy weather to interfere with grasshopper activity. A single season of such weather is not adequate to stimulate massive population increase; rather, two to three years of drought usually precede grasshopper outbreaks. Warm winter temperatures also seem to be beneficial, because fewer overwintering nymphs and adults die.

Food is necessary for grasshopper success, and optimal weather alone in the absence of adequate food supply will prove insufficient for rapid grasshopper population growth. For outbreaks to occur, both requisites must be satisfied. Thus, some precipitation must be present at the appropriate time to stimulate plant growth, but an overabundance results in too much cloud cover. Optimal weather conditions and food supply rarely coincide in Florida, and so outbreaks are infrequent.

An example of a grasshopper outbreak in central Florida in the early 1990s illustrates the importance of the interaction of weather and food. Abnormally cold winters killed most of the citrus grown north of Orlando during the mid to late 1980s. Extensive acreage was abandoned as citrus growers moved their production farther south. This set the stage for an increase in grasshopper food supply, as untended land produced luxuriant growth of mixed grasses and weeds. Coincidentally, most of Florida experienced about five years of drought. This was a period when even the Everglades, normally a very wet environment, became very dry. The availability of extensive amounts of low-growing, mixed grass and broadleaf weed herbage in a dry, sunny environment stimulated massive grasshopper population increase in what was formerly the northern citrus-growing area. Although the weather pattern was widespread in Florida, the grasshopper problem was restricted to areas that contained large quantities of the appropriate food.

Grasshopper Management

Once grasshoppers become abundant and damaging there are few options for suppression other than chemical insecticides. People often ask about the potential for biological suppression of grasshoppers. Yes, wild birds readily consume vast quantities of grasshoppers, but this is a natural check on their populations that will not completely eliminate them. Domestic fowl, particularly turkeys, are also very fond of grasshoppers but few homeowners have the desire or the resources to keep turkeys. There also are grasshopper disease agents under investigation, and even some that are sold commercially, but so far none has been shown to provide adequate suppression. So biological control remains a promising area for research, and we continue to hope for development of an effective product, but thus far there is nothing practical. For some people, neem products are attractive. Neem products are derived from the oil of seeds of a southeast Asian tree and, when applied to plants, make the plants unpalatable to insects, thereby reducing damage. Also, if applied to grasshopper nymphs, neem can act as a growth regulator, disrupting the normal growth and development, and sometimes resulting in death or sterilization of grasshoppers. Although neem products are chemicals, many people take comfort in knowing that they are derived from plants, and therefore somewhat "natural." Like many natural controls, effectiveness is not always consistent.

For many people, physical barriers can provide some protection from damage. It is possible to screen or cover valuable plants with netting, floating row covers, or similar material to deny grasshoppers access to susceptible plants. This is suitable for small gardens, and is even applied commercially for ornamental plant production, wherein shade houses are sealed tightly to deny access by grasshoppers. The potential for this approach is limited in scale due to the cost. For lubber grasshopper, a flightless species, physical barriers in the form of a ditch with steep sides, or a short metal or plastic wall, can be an effective impediment to grasshopper dispersal. If such a wall is contemplated, however, consider that grasshoppers can ascend vertical surfaces with amazing agility, so the tops of barriers should end in a 45-degree angle, forcing the insects to fall back.

The best approach to grasshopper management is to strive to avoid problem populations through cultural manipulation of their habitat (see below). This is not always possible due to the highly dispersive nature of some species, so chemical insecticides commonly are used.

Cultural Approaches to Grasshopper Management

As discussed elsewhere (see "Grasshopper Habitats"), the habitats most favorable for grasshopper population growth and survival are open, sunny habitats containing mixed, early- to mid-successional plants. Land with trees providing moderate to deep shade rarely produces large numbers of grasshoppers. Also, land that is kept mowed mechanically or by livestock grazing tends not to produce grasshoppers unless grass pasture is damaged by overgrazing and broadleaf weeds invade.

Crops planted near habitats containing dense, mixed, early- and mid-successional plants may experience invasion by grasshoppers. This has been particularly evident in north and central Florida with respect to slash pine plantations. Young pine trees, once planted, are largely ignored until they are partly mature and the stand requires thinning. In the early years following pine seedling establishment, mixed grasses, broadleaf weeds, and small shrubs flourish, but they usually are ignored because the fast-growing pines eventually outgrow the weeds. This results in excellent habitat for grasshoppers that can then spread to adjacent crops (as the pine stand matures, of course, it shades out undergrowth and is poor habitat for grasshoppers). If crops are planted near young pine plantations, it is important to recognize this potential source of grasshoppers. If grasshoppers attain high densities and appear to be threatening it may be desirable to mow or plow between the pines, depriving grasshoppers of habitat and disrupting their biology. Alternatively, insecticide can be applied to the pine plantation, or perhaps only the border of the plantation, thereby establishing an insect-free buffer between the source of grasshoppers and the susceptible crops.

There is one important exception to the pattern described above, and this involves the eastern lubber grasshopper, *Romalea microptera*. This unusual, flightless grasshopper is one of the more common pests in Florida, and is found damaging vegetable gardens, ornamental plantings, and citrus groves in a variety of habitats. However, despite its widespread occurrence in the late nymphal and adult stages, it seems to come every year from low, moist, dense habitat, including areas that include considerable tree overstory.

Insecticides for Suppression of Grasshoppers

Chemical insecticides can be applied in liquid form, by application directly to the grasshoppers or to the plants they will walk or feed on. Insecticides can also be applied to bran flakes, and distributed as a bait. Both approaches, but particularly the latter, are difficult to use in Florida due to dense vegetation. If insecticides are to be used, it is advisable to apply the chemicals when the grasshoppers are young. Small insects are much easier to kill than large, and grasshoppers are notoriously difficult to kill under any conditions. Also, because the grasshoppers usually develop in surrounding vegetation it is usually best to take the "battle" there, and apply insecticides to the young grasshoppers before they disperse into crops and cause damage.

What Is a Species?

Often it is difficult to decide whether populations differing slightly from one another should be regarded as different species, or simply as varieties or subspecies. This is especially true with populations that are geographically isolated.

Most biologists define species as groups of actually or potentially interbreeding populations that are reproductively isolated from other groups. Thus, if populations of grasshopper *X. a* in southeast Florida breed successfully with populations in southwest Florida and central Florida, we consider them to be the same species. If species *X. a* from central Florida does not successfully breed with grasshoppers from north Florida in the regions where both occur we consider the northern populations to be a separate species, perhaps designated as *X. b*. But what do we make of grasshoppers that occur only in the western Panhandle area of Florida if they do not breed successfully with our northern *X. b* grasshoppers? Are they another species, perhaps *X. c*, or could they be geographically separate populations of our southern and central Florida species, *X. a*? Because the geographically separate populations do not have the opportunity to interbreed we cannot easily test the definition of a species.

Florida has many grasshoppers with strong powers of flight, and broad dietary habits, allowing them to move freely across the landscape. When such grasshoppers segregate into separate populations, and attain the different appearances or behaviors that normally accompany reproductive segregation, we have no difficulty designating them as separate species. However, Florida also has many shortwinged, flightless grasshoppers that dwell on isolated sand ridges. During ancient times when the sea level was higher, these ridges were islands, and even now are "biological islands," with unique flora. Thus, there have been, and remain, isolated populations of grasshoppers incapable of reproduction because the populations of one "island" do not come into contact with the population of other "islands." But are they truly reproductively isolated? Have they diverged sufficiently to be considered true species? Often it is difficult to know with certainty whether different populations constitute different species. However, if they differ greatly in appearance or behavior, especially if the differences are associated with reproductive structures or mating behavior, we tend to consider them separate species. There is some disagreement among scientists, of course, over whether certain populations are sufficiently different to be considered separate species. In this manual we have taken a conservative approach, consolidating some very similar populations that others have considered to be separate species into a smaller number of species.

Why is it sometimes difficult to determine if populations represent different species? As previously mentioned, speciation is based on behavior, physiology, and genetics, as well as appearance, or morphology. Only morphology is relatively easy to assess and to portray to others. Therefore, most taxonomists depend on morphological characters to distinguish among species. Unfortunately, there can be considerable variation in morphology within even a small geographic area, and immense variation over the entire range of an insect. Many scientists have described grasshoppers as new species only to discover later that the new "species" was only a morphological variant of another species. Grasshoppers exhibit considerable variation in color and wing length, accounting for many of the erroneous species descriptions. Often the sexual organs are used to distinguish among species, because these characters relate directly to reproductive compatibility and population genetics. However, even here it is sometimes difficult to know how much difference in genitalia constitutes enough to cause physical incompatibility and result in genetic isolation.

Exact determination of species numbers is probably a fruitless enterprise, as speciation is a dynamic phenomenon, with new species slowly evolving in some areas, and extinction probably occurring in others. Particularly in the case of Florida, where land conversion and natural habitat destruction are frequent, extinction is a real possibility for some species of grasshoppers. Perhaps the best example of this is *Melanoplus puer* and its closely related species. *Melanoplus puer* has several races or subspecies that differ only slightly, but most are separated geographically and seem to be in the process of evolving into separate species. Some of *M. puer*'s close relatives are only marginally different from it, and likely evolved in relatively recent times. Also, such species as *Melanoplus adelogyrus, M. apalachicolae, M. gurneyi, M. indicifer, M. pygmaeus* and *M. withlacoocheensis* apparently occur in small areas. *Melanoplus gurneyi* occurs in a fairly small area in the western Panhandle area of Florida, and *M. indicifer* in an even smaller area near Jupiter on the east coast of Florida. If extensive development occurs in these areas the probability of such species surviving is remote.

Collecting and Preserving Grasshoppers

Why Collect Grasshoppers?

Grasshoppers are among the best insects to collect. You do not have to travel far or to exotic habitats to collect some, and often they are numerous. They commonly are large, at least by insect standards, which makes them relatively easy to handle and identify. Identification ranges from quite simple to requiring close, detailed examination, so there are various levels of challenge. They also vary in difficulty of collection; some species are easy to obtain whereas others require considerable searching or travel to a specific location in the state. Grasshoppers are inherently interesting, and often strikingly beautiful if you take the time to look closely. Few people have seen the myriad of hues on the hind wings of bandwinged grasshoppers. Fewer yet have examined the inside of the hind legs of these species for the striking array of colors, including bands of yellow, orange, red, blue, and black. Finally, there is considerable physical and intellectual challenge in netting these cunning species. The grasshoppers that respond to your net by flying 100 meters away or 10 meters up into a tree or resting on emergent vegetation in a pond frequented by alligators will quickly earn your respect!

Collection

Grasshoppers are easily collected and preserved. An insect net is usually required for collection, especially for the bandwinged and *Schistocerca* species. Other than a net, no special equipment is required. Because immature stages are frequently collected, have a cage or other container available to rear the nymphs to the adult stage.

Grasshoppers are usually easy to culture if you are only trying to obtain the adult stage. Grasshoppers like to be kept hot and require a fresh food supply. The principal trick is to be sure that the correct food is supplied. It is always best to feed grasshoppers with whatever they are feeding on when they are collected. Grass-feeding species normally require grasses. If in doubt, try St. Augustine grass. Species preferring broadleaf plants may not accept grass. Romaine lettuce, but not iceberg lettuce, is often a good substitute for broadleaf plants. Replace green food every two days.

To culture grasshoppers through a complete generation is a more formidable task. Grasshoppers must be provided with a place to deposit their eggs, usually a cup of soil. The eggs must be held at warm temperature for a month or more, then usually benefit from 2–3 months of refrigeration before being brought out into a warm environment for hatching. Eggs are tolerant of variable moisture conditions, but cannot be allowed to dry out completely. Grasshoppers usually benefit from being able to regulate their own body temperature by moving close to, or away from, a heat source. The safest way to accomplish this is to provide an electric light bulb in the cage or nearby.

It is easy to find habitats suitable for grasshopper collecting. In fact, this is one of the advantages of collecting grasshoppers; nearly all habitat types support interesting and unique species. It is possible to stop along the side of any road in Florida that is bounded by grass, weeds, and low-growing or open vegetation, and find large numbers of grasshoppers. Collecting can be done throughout the year, but the best period to collect is April to

November. Most landowners are extremely cooperative and freely allow insect collecting, but if the landowner can be found it is best to inquire before collecting. National and state parks usually prohibit removal of anything from the property, so inquire in advance and request a permit if you want to collect in such habitats.

Preservation of Specimens

Grasshoppers normally are preserved by killing, pinning, and drying. Grasshoppers can be killed by freezing or with chemicals. The easiest, safest, and perhaps most humane technique is to place insects into a freezer for several hours. They can also be killed by exposing them to a small amount of toxic fumigant such as ethyl acetate, the major component in nail polish remover, in a tightly sealed container. Toxicants are usually used in conjunction with a specially prepared killing jar, which is easily made by pouring plaster of paris (in a 1:1 ratio of plaster to water) into the bottom of a wide-mouth jar to a depth of about 1/2 or 3/4 inch. Pour ethyl acetate onto the dried plaster, allow several minutes for the liquid to be absorbed, then pour off any remaining liquid. The jar will produce toxic fumes for several days if the lid is kept tightly sealed.

Usually it is not desirable to kill nymphs because they lack the characters needed for identification. Also, owing to their soft bodies they do not

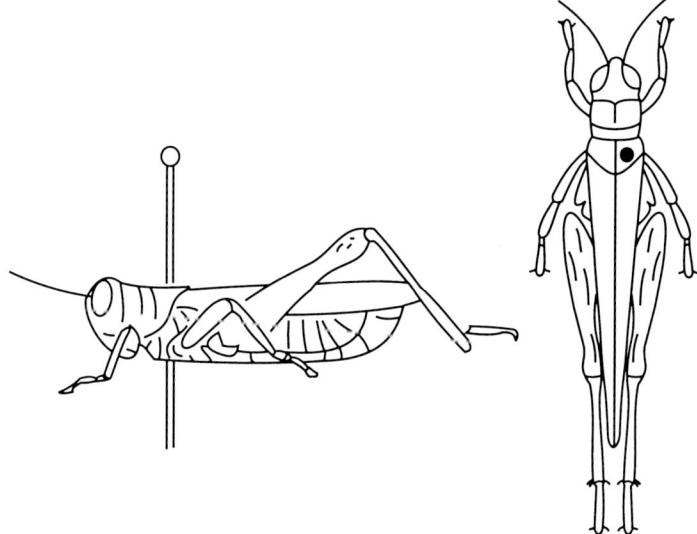

Figure 8. Pinned grasshoppers showing proper placement of pin.

preserve well in a dry state. They are best placed in alcohol to prevent excessive distortion.

To mount adult grasshoppers on a pin, insert a pin into the dorsal, or upper, surface, with the point protruding from the ventral surface. The preferred location for pinning is usually the rear of the prothorax, and to the right of the midline. The grasshopper is pushed up on the pin so that not only the end, but a small amount of the shaft is protruding. This gives ample room to pick up the dead grasshopper without touching the insect's body. Below the grasshopper body, collection data are provided via a label. This is accomplished by writing or printing data on stiff paper, and cutting the label to a small rectangle. A pinning block often is used to align the insect body and label(s) to standard heights. Data that should be included on the label include the date of collection, place of collection, and collector's name. Ecological data such as habitat or host plant may also be included. Pins vary in size and quality. It is highly desirable to use rust-proof insect pins. Insect pins are longer and sharper than standard pins, allowing attachment of labels and easy mounting. Insect pins and all other collection and preservation equipment are available from biological supply houses.

To fully appreciate the beauty of the bandwinged species, and to assist in identification, spread at least one forewing and hind wing. The usual procedure is to spread the left forewing perpendicular to the grasshopper body. Similarly, the leading edge of the hind wing is spread perpendicular; this results in full extension of the remainder of the hind wing. Species other than bandwinged grasshoppers are rarely spread, though if it is done the spread wing may aid identification.

To properly spread the grasshopper's wings, some support is needed to keep the wings elevated and flat. A spreading board is usually used to provide wing support. A spreading board consisting of plastic foam or another suitable pinning surface should have a strip of similar material glued on part of the board, so that one surface is higher than the other. Thus, the lower pinning board is used to support the grasshopper body on its pin, and the elevated portion is used to support the wing. Strips of paper and pins are used to hold the wing in place, with the weight of the paper holding the wing and the pins holding down their ends and not piercing the wing. Whether or not the grasshopper's wings are spread,

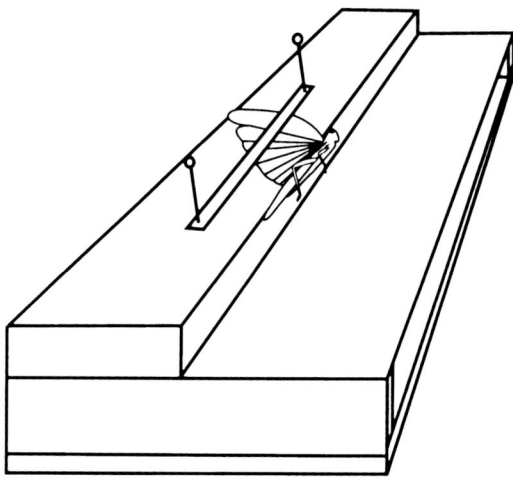

Figure 9. Grasshopper with wing on spreading board.

grasshoppers must be dried to aid preservation. Specimens will dry naturally in a few days, and pinning them to a vertical surface as they dry will prevent the abdomen from sagging. Drying can be hastened by placing the pinned insect, often with its wings spread and abdomen supported on two crossed pins, in an oven at very low temperature until the subject is dry and stiff. Once dried, the wings, antennae, and legs cannot be moved without breaking, so it is important to get the body parts aligned before drying.

The color of a grasshopper tends to fade as the insect dies. This is difficult to prevent. Much of the discoloration is due to the accumulation of body oils at the surface of the body. The oil can be extracted, preventing some of the color change. To extract oils, place the dried insect on its pin in a bath of acetone. Usually a few hours is adequate; prolonged extraction causes the insect body to bleach to a light color.

If you are going to make an insect collection you need suitable storage. Storage requires nothing more than a tight box with pinning material such as plastic foam or balsa wood in the bottom. However, it is imperative that the box be tight, or carpet beetles and cockroaches will gain access and devour the pinned insects. To help prevent damage to specimens, you can place moth balls or moth crystals in the box with the specimens. This will kill any insects that gain access, particularly ants, cockroaches, and book lice.

2

Quick Guide to Common Florida Grasshoppers

 This guide is designed to assist in the identification of common grasshoppers in Florida. The guide will allow you to quickly eliminate most species from consideration, allowing you to focus on the most probable identifications. It does not contain most of the uncommon short-winged *Melanoplus* species, but all other species are represented. The *Melanoplus* species are best identified by examination of terminal abdominal structures of males; diagrams of these are provided in the species descriptions. To obtain accurate identifications, the guide should be used in conjunction with the descriptions, diagrams, and photographs. This guide will be less reliable in nearby states where other species occur that are not found in Florida.

Only a limited amount of terminology is needed to understand the characters used to identify grasshoppers. Diagrams of grasshoppers are shown in figures 1–4, and terms are defined in the glossary. The key characters are the occurrence and length of wings in the adult stage; the overall size of the adult; the body color, especially pigmentation on the wings and legs; the orientation of the face (slanted or vertical); and the shape of the male genital structures. The sex of the insect is most easily determined by examining the tip of the abdomen; the female has pointed structures that open (fig. 3). Immature grasshoppers are usually recognized by poorly developed wings or by underdeveloped genital structures.

WINGS LACKING; *or apparently no wings*

Size small (12–22 mm in length), color gold or brown
    Gymnoscirtetes morsei, G. pusillus
Size medium (15–33 mm), color green
    Aptenopedes aptera, A. sphenarioides

WING LENGTH SHORT; *wings distinct but less than, or equal to, length of pronotum*

Body form exceptionally long and narrow
    Achurum carinatum
Body with a bold white stripe on top of pronotum and abdomen
    Hesperotettix osceola, Eritettix obscurus
Body color uniformly bright green with, at most, a weak red stripe on top of pronotum
    Hesperotettix floridensis
Body color iridescent yellowish, gold, or brown
    Eotettix species
Body color indistinct brownish, reddish, or grayish, and with black stripe on side of pronotum
    Male with distinct conical structure (pallium) pointing upward near tip of abdomen
        Melanoplus rotundipennis, M. withlacoocheensis
    Male without distinct conical structure
        Several short-winged *Melanoplus* species, usually uncommon

WING LENGTH INTERMEDIATE; *wings appreciably longer than pronotum but not reaching tip of abdomen*

Size small (16–28 mm), color usually grass-green
    Stripe absent from top of pronotum
    Dichromorpha elegans, D. viridis

    Stripe present on top of pronotum
    Hesperotettix viridis

Size medium (22–40 mm), color usually olive green or brownish
  Melanoplus querneus
Size large (43–70 mm), forewing color some combination of black,
yellow, and reddish
  Romalea microptera

WING LENGTH LONG; *wings nearly as long as abdomen or extending beyond tip*

Hind wings distinctly pigmented, usually brightly colored with black
band across them
  Hind wing orangish or pinkish
    Black band wide, about 1/3 the width of the wing, and crossing
    near the center of the wing
      Psidinia fenestralis, Spharagemon marmorata
    Black band not wide, about 1/4 the width of the wing or less, and
    not crossing centrally
      Hippiscus ocelote, Pardalophora phoenicoptera
  Hind wing yellow
    Hind wing lemon yellow at point of attachment, wing tip usually
    cloudy
      Arphia species, *Spharagemon marmorata*
    Hind wing pale yellow at point of attachment, wing tip usually
    transparent
      Hippiscus ocelote, Spharagemon bolli, S. crepitans,
      S. cristatum, Trimerotropis maritima
  Hind wing black, with yellowish edge
    Dissosteira carolina
  Hind wing largely transparent, with diffuse blackish area centrally
    Chortophaga australior
Hind wings not distinctly pigmented, usually transparent except for
wing veins
  Face strongly slanted; spine present or absent between front legs
    Tips of forewings sharply pointed; spine present between front legs
    Leptysma marginicollis, Stenacris vitreipennis

    Tips of forewings flattened, but forming sharp angle; spine absent
    between front legs
    Metaleptea brevicornis

Tips of forewings rounded; spine absent between front legs
   Antennae markedly flattened and sword-shaped
   Mermiria species

   Antennae not markedly flattened and sword-shaped
   Sides of dorsal surface of prothorax well marked with white lines
   Orphulella pelidna, Syrbula admirabilis

   Sides of dorsal surface of prothorax not marked with white line
   Amblytropidia mysteca, Dichromorpha elegans, D. viridis

Face not strongly slanted; spine present between front legs
   Cerci of males broad, flat, with tip wider than base
      Melanoplus furcatus, M. keeleri, M. punctulatus, M. symmetricus
   Cerci of males with tip width narrower than base width
      Cerci spoon-shaped at tip
      Paroxya species, *Melanoplus impudicus*

      Cerci with blunt or rounded tip, but not spoon-shaped
      Hesperotettix viridis, Melanoplus propinquus, M. sanguinipes

   Cerci of males with tip width about same as base width
      Cerci spoon-shaped at tip
      Paroxya species, *Melanoplus bispinosus*

      Cerci about equal in width throughout and flattened at tip
      Schistocerca species

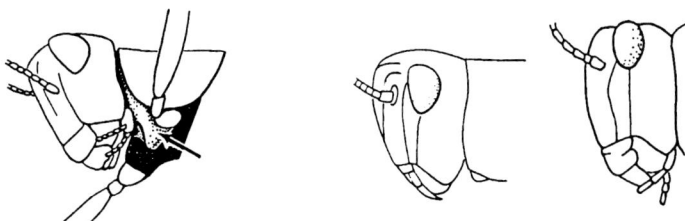

Lower (ventral) view of
grasshopper showing spine
between base of front legs

Examples of grasshoppers
with face not strongly slanted

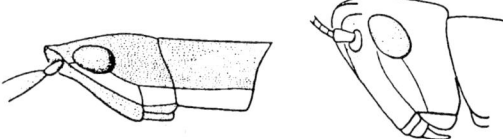

Examples of grasshoppers
with a strongly slanted face

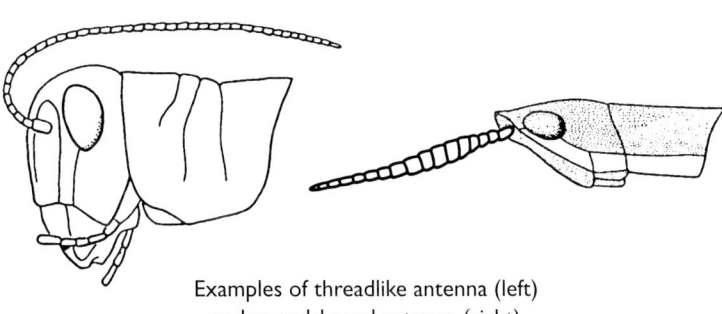

Examples of threadlike antenna (left)
and swordshaped antenna (right)

Figure 10. Identifying characteristics of grasshoppers.

3

Profile of Florida Grasshopper Species

Stridulating Slantfaced Grasshoppers

Subfamily GOMPHOCERINAE

 Grasshoppers in this subfamily tend to have slender bodies and long, slender legs. Their heads are elongate and often cone-shaped, usually having a highly slanted face. They usually lack a spine between the front legs (the prosternal spine), as is found in the lubber grasshoppers (subfamily Romaleinae) and the spurthroated grasshoppers (subfamily Cyrtacanthacridinae). Gomphocerine grasshoppers tend to be green or brown; sometimes distinctly brown or green forms occur within the same species. The hind wings are not colorful.

Gomphocerines often have relatively short wings, so they are not capable of sustained flight. When disturbed these grasshoppers leap and use their wings, but their wings often do little more than increase the distance jumped. They do not make sounds during flight, called crepitation, as occurs in the bandwinged grasshoppers (subfamily Oedipodinae). This does not mean that these grasshoppers are silent, because they can make noise by rubbing the inner surface of the hind femur on the edges of the forewing. They create this sound, called stridulation, while resting, not while flying. Because the males of this subfamily usually have a row of stridulatory pegs on the inner surface of the hind femora, they are also known as tooth-legged grasshoppers.

The habitat of gomphocerines tends to be tall grasses in open fields. The form and color of many species allow them to blend in with stems and blades of grass, making them difficult to detect until they move. Most species feed predominantly on grasses.

There are 10 species in 7 genera of Gomphocerinae in Florida:

Achurum
    A. carinatum (F. Walker)
Amblytropidia
    A. mysteca (Saussure)
Dichromorpha
    D. elegans (Morse)
    D. viridis (Scudder)
Eritettix
    E. obscurus (Scudder)
Mermiria
    M. bivittata (Serville)
    M. intertexta Scudder
    M. picta (F. Walker)
Orphulella
    O. pelidna (Burmeister)
Syrbula
    S. admirabilis (Uhler)

Achurum carinatum (F. Walker)

Longheaded toothpick grasshopper

Plate 20. Longheaded toothpick grasshopper (male).

Identification. The common name of this slender grasshopper accurately describes its general body form. It is pale brown or grayish brown, often with the forewings and legs partly green. Thus, it easily blends in with grasses and pine needles, and is difficult to detect. The forewings of this flightless species are small, averaging about the length of the head in north

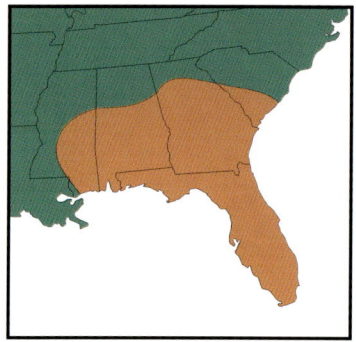

Florida and shorter than the head in south Florida. The face is extremely slanted, the antennae large and sword-shaped. Some individuals bear numerous black dots. The length of males is 24–36 mm; females are 33–40 mm.

Similar Species. The short wings and absence of a spine between the front legs serve to differentiate the longheaded toothpick grasshopper from the similar *Leptysma marginicollis* and *Stenacris vitreipennis,* which are in the subfamily Cyrtacanthacridinae.

Distribution and Ecology. The longheaded toothpick grasshopper is found throughout Florida, and the southeastern states from South Carolina to Mississippi. It can be found throughout the year, with nymphs overwintering in north Florida and adults farther south. It is commonly found in grass of open woodlands such as high pine and pine flatwoods, and areas with tall grasses such as old fields and pond margins.

Amblytropidia mysteca
(Saussure)

Brown winter grasshopper

Plate 21. Brown winter grasshopper (male).

Identification. This yellowish brown to brownish black grasshopper is heavybodied in form. Although the face is strongly slanted, the top of the head is broadly rounded. The antennae are relatively short. The top of the head, thorax, and sometimes a portion of the forewings are yellowish brown or gold. The junction of the top and side surfaces often is marked by a narrow but distinct line. The tips of the forewings are darker brown or blackish. The hind femora are light brown or gold, sometimes with a black line running their lengths. The hind tibiae are light brown where they join the body and brownish black at the ends. The length of this insect is 19–24 mm in males and 24–30 mm in females.

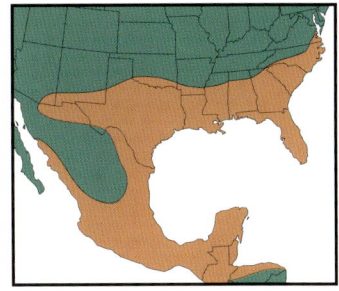

Distribution and Ecology. The brown winter grasshopper is found throughout Florida and occurs widely in the southern United States from North Carolina to Arizona. It can be found in Florida during the winter in both nymphal and adult stages, but also is collected through most

Map 2. Brown winter grasshopper distribution.

of the summer. It occurs among short and moderately high grasses, usually in open woodlands. When disturbed *A. mysteca* flies short distances, dives into vegetation, and burrows out of sight among the foliage and debris. This is an unusual and easily recognizable behavior that aids in field identification.

Dichromorpha elegans (Morse)

Elegant grasshopper

Plate 22. Elegant grasshopper (female).

Identification. This grass-green or brownish green grasshopper is attractive, but hardly deserves its common name; there are many other more elegant species. It has a slanted face, but broadly rounded, slightly enlarged head. It usually is marked by a narrow black line extending from behind the eye, across the prothorax, and onto the forewings. The forewings are variable in length, ranging from about one half the length of the abdomen to the tip of the abdomen. The males are much smaller and more slender than the females. The hind tibiae are brownish. The length of this grasshopper is 17–21 mm in males and 19–28 mm in females.

Similar Species. This species is easily confused with *Dichromorpha viridis.* Close examination of the pronotum will differentiate the two species: *Dichromorpha elegans* has a single, narrow, line-like groove that runs across the pronotum, crossing ridges on the sides and top. *Dichromorpha viridis*

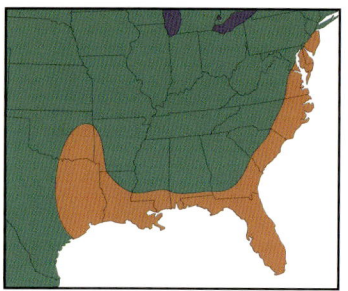

Map 3. Elegant grasshopper distribution.

has, in addition, another crevice that bisects the side ridges but not the one in the middle. *Dichromorpha elegans* also has a larger head than *D. viridis.*

Distribution and Ecology. The elegant grasshopper is found throughout Florida in grassy, moist areas such as freshwater swamps and salt marshes. It does not inhabit the dryer areas where *D. viridis* is found. It also occurs elsewhere in the eastern and southern United States along both the Atlantic and Gulf coasts.

Dichromorpha viridis (Scudder)

Shortwinged green grasshopper

Plate 23. Shortwinged green grasshopper (male).

Identification. Although this common name may not seem unique or descriptive, it actually is appropriate, because this is the most abundant short-winged green grasshopper species in Florida. The top and side surfaces of males of this species sometimes are contrasting colors. The common forms are a green upper surface and pale or dark brown sides, or light brown upper surface and dark brown sides. Females are uniformly colored, but may be either green or brown. The forewings of this species are, as the common name suggests, usually short, but occasional longwinged individuals occur. In males, the forewings often are about three-fourths the length of the abdomen, while in females the forewings often extend just half the length of the abdomen. The males measure 14–17 mm in length, the females 23–27 mm.

Plate 24. Shortwinged green grasshopper (female).

Similar Species. This species is very similar in form to *Dichromorpha elegans*. As previously noted, *D. viridis* can be distinguished by the presence of two crevices or cuts in the ridges on the sides of the pronotum. Also, *D. elegans* has a larger head than *D. viridis,* but of course this character is useful only if individuals of both species are available to compare.

Distribution and Ecology. This is a common species in grassy areas, including edges of ponds and woods, low areas of pastures, and along roadsides. It also feeds readily in improved pastures and on lawn grasses, which accounts for its wide distribution and abundance. In many grassy habitats it is the most abundant grasshopper. Shortwinged green grasshopper can be found throughout the year in Florida, although it is infrequent in north Florida during the winter. Males are reported to stridulate.

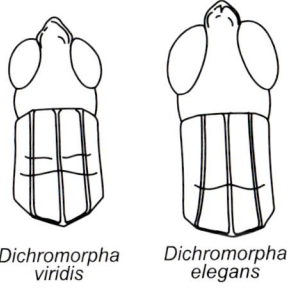

Dichromorpha viridis   *Dichromorpha elegans*

Figure 11. Two crevices, or cuts, present *(left)* and absent *(right)* in the side ridges of the pronotum.

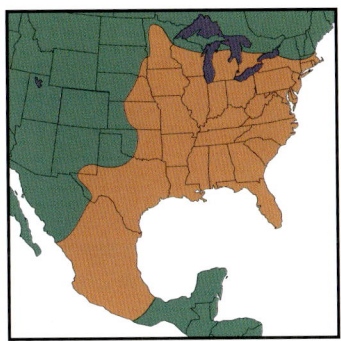

Map 4. Shortwinged green grasshopper distribution.

Eritettix obscurus (Scudder)

Obscure slantfaced grasshopper

Plate 25. Obscure slantfaced grasshopper (female).

Identification. This small brownish, flightless grasshopper is usually distinguishable by the broad whitish or yellowish stripe that extends from the top of the head to the tip of the abdomen. Unfortunately, this stripe is sometimes absent, making recognition more difficult. The forewings are always shortened, covering about one-half to three-fourths of the abdomen. The antennae are slightly flattened and sword-shaped. The face is very slanted. The pronotum bears a small ridge along each side, and the ridges are constricted, or pinched, near the middle of the pronotum. These lateral ridges may be black or white. Males measure 13–15 mm in length, females 21–24 mm.

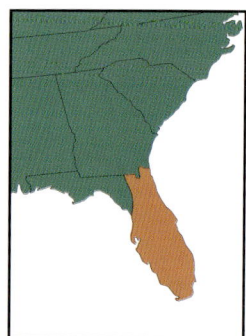

Map 5. Obscure slantfaced grasshopper distribution.

Distribution and Ecology. The obscure slantface grasshopper occurs only in Florida, and widely in the peninsular region from Live Oak to Miami. Its preferred habitat is scrub oak forest, where it occurs among wire grass and low-growing oak. The obscure slantface grasshopper is not uncommon in such habitat but never is abundant.

Mermiria bivittata (Serville)

Twostriped mermiria grasshopper

Identification. This narrowbodied species is quite large. Though variable in color, it is marked with a dark stripe originating behind the eye and running across the pronotum. The stripe extends weakly onto the fore-

Plate 26. Twostriped mermiria grasshopper (female).

wings, where a narrow white streak also may be found where the wing joins the body. Generally the body is brownish or greenish on the top and yellow underneath. The face is strongly slanted. The antennae are sword-shaped. The hind tibiae are reddish. Body length is 28–38 mm in males and 39–56 mm in females.

Similar Species. Twostriped mermiria grasshopper is separated from *Mermiria intertexta* by the lack of a dark stripe running from the tip of the head to the rear edge of the pronotum; this stripe is apparent in *M. intertexta*. Twostriped mermiria grasshopper is distinguished from *Mermiria picta* by the absence of distinct ridges on the sides of the pronotum; the lateral ridges are evident in *M. picta*.

Distribution and Ecology. The twostriped grasshopper is found nearly everywhere in the United States except the northwestern and northeastern states. It also is known in southern Canada and northern Mexico. It is most common in the Great Plains region, however. In Florida it is known in the northernmost counties, but is not generally common. Twostriped mermiria grasshopper inhabits areas of tall grass, including coastal salt marsh habitats. It feeds exclusively on grasses.

Mermiria picta

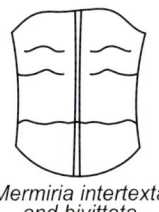

Mermiria intertexta and bivittata

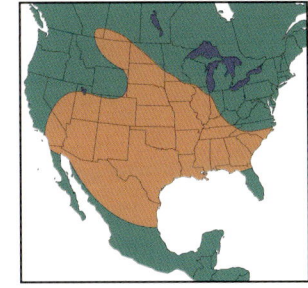

Figure 12. Side ridges present on pronotum *(left)* and absent *(right)*.

Map 6. Twostriped mermiria grasshopper distribution.

Mermiria intertexta
Scudder

Eastern mermiria
grasshopper

Plate 27. Eastern mermiria grasshopper (female).

Identification. This species is long and narrow in general appearance, and fairly large. The face is strongly slanted, and the antennae sword-shaped. The general color is yellowish or greenish, but a reddish or dark brown stripe often is present on the upper surface, especially in males, from the tip of the head to the rear edge of the pronotum. Another distinct dark brown or black stripe extends from the rear edge of the eye onto the base of the front wings, and merges into the brown forewings. The stripe where the forewings join the body contains a narrow white streak. The long, thin hind tibiae are reddish. The males measure 32–38 mm in length, the females 33–58 mm.

Similar Species. The narrow white streak found at the base of the forewing is lacking in the similar *Mermiria picta*. Also serving to distinguish *M. intertexta* from *M. picta,* but much more difficult to see, is the absence of distinct ridges on the sides of the pronotum; these lateral ridges are

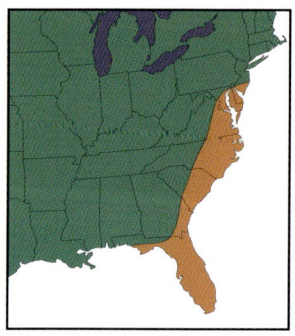

Map 7. Eastern mermiria grasshopper distribution.

present in *M. picta*. The stripe down the middle of the head and thorax of *M. intertexta* is absent from *M. bivittata*.

Distribution and Ecology. The eastern mermiria grasshopper inhabits the coastal region of the eastern United States from New Jersey to Florida. In Florida, it occurs throughout the state, usually in tall grasses and in wet habitats. Despite its ability to blend in well with its grassy environment, *M. intertexta* flies freely when disturbed. It is not a strong flier.

Mermiria picta
(F. Walker)

Lively mermiria
grasshopper

Plate 28. Lively mermiria grasshopper (male).

Identification. This large, thin species greatly resembles *Mermiria inter-texta*: long and narrow in general appearance, and greenish or brownish in general color. The face is strongly slanted, and the antennae sword-shaped. A reddish or dark brown stripe often is present on the upper surface, especially in males, from the tip of the head to the rear edge of the pronotum. Another distinct, dark brown or black stripe extends from the rear edge of the eye onto the base of the front wings, and merges into the brown forewings. The antennae of lively mermiria grasshopper are sword-shaped. The hind tibiae are reddish. The males measure 28–41 mm in length, the females 41–57 mm.

Similar Species. This species does not have a narrow white streak at the base of the forewing, a character that is found in the similar *Mermiria intertexta,* and usually in *M. bivittata.* Serving to distinguish *M. picta* from both *M. intertexta* and *M. bivittata,* but much more difficult to see, is the presence of distinct ridges on the sides of the upper surface of the pronotum in *M. picta.* Also, the stripe found on the upper surface of the head and thorax of lively mermiria is normally absent from *M. bivittata.*

Distribution and Ecology. The lively mermiria grasshopper is found throughout the eastern United States north to Virginia and South Dakota, and west to Arizona. It also is known in Mexico. In Florida it occurs widely in habitats containing tall grasses, including wooded environments. It tends to inhabit drier areas than *M. intertexta,* and is less frequently encountered.

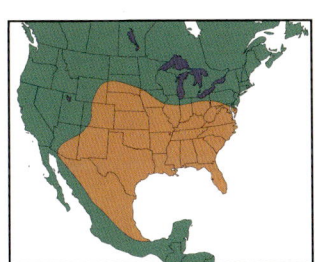

Map 8. Lively mermiria grasshopper distribution.

Orphulella pelidna
(Burmeister)

Spottedwinged
grasshopper

Plate 29. Spottedwinged grasshopper (female).

Identification. This slender species is variable and indistinct in appearance. Under most conditions, spottedwinged grasshopper is brown or green and bears both black and white accents. Large black triangular marks are found dorsally along the upper surface at the back edge of the pronotum. The ridges along the sides of the upper surface of the pronotum are markedly compressed, coming closer together near the midpoint of the pronotum. A series of small, dark, rectangular spots is present along the center of the forewings, and is the basis for the common name of this species, but numerous speckles are also generally present. A broad dark stripe usually extends from the back of the eye to the base of the forewing. In some coastal locations larger forms appear. They may be completely green, or bear only some of the stripes and marks mentioned previously. On dark soils or in burned forests, blackish forms may occur. Also, although the forewings normally extend well beyond the tip of the abdomen, individuals with shorter wings are sometimes observed. The hind tibiae are usually brown, but sometimes bluish. The antennal segments are not markedly flattened, appearing threadlike. The males measure 18–25 mm in length, the females 18–28 mm.

Similar Species. Although this species superficially resembles *Syrbula admirabilis,* it usually can be separated easily based on its overall

Plate 30. Spottedwinged grasshopper (female).

smaller size. The strongly compressed lateral ridges also are diagnostic, although in *S. admirabilis* they are slightly compressed. Also, the forewings are usually spotted and speckled, characteristics usually absent from *S. admirabilis*.

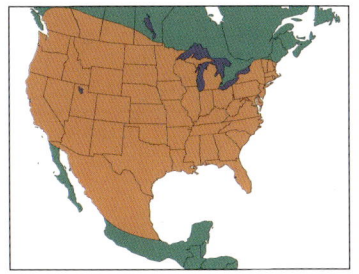

Map 9. Spottedwinged grasshopper distribution.

Distribution and Ecology. This is the most widespread grasshopper in Florida. It is found in all habitats except those that are very shaded. Despite its adaptable nature, it rarely is numerous in any habitat, or at any location. When disturbed it flies swiftly, but fairly short distances, before diving to the soil or vegetation and seeking shelter. Upon landing it often runs a short distance to hide. In addition to being found throughout Florida, *O. pelidna* occurs throughout the United States, and even into southern Canada and northern Mexico.

Syrbula admirabilis (Uhler)

Handsome grasshopper

Plate 31. Handsome grasshopper (male).

Identification. This slender species deserves its common name; it is a strikingly attractive insect. The face is quite slanted, the hind legs especially long and slender. The antennae, though generally threadlike, in males are slightly expanded at the tip. The hind tibiae are brownish. The pronotum bears small ridges on the sides, which are marked by white stripes and pinched slightly near the middle of the pronotum. A broad brown stripe usually extends on the upper surface from the front of the head to the rear edge of the pronotum. The general body color ranges from mostly brown to mostly green, but some individuals tend toward blackish, especially males. The

Plate 32. Handsome grasshopper (female).

most distinctive feature is the pattern on the forewings. The leading edge of the forewing (undersurface when wings are closed) is green or grayish, whereas the trailing edge is brown to black. These contrasting colors meet in a wavy, or crenulate, pattern that immediately distinguishes most individuals. Some males, however, have the forewings almost entirely dark. The sexes differ markedly in size. Males measure about 22–28 mm in length, females 35–42 mm.

Similar Species. Although usually distinct, this species sometimes resembles *Orphulella pelidna.* Usually it can be separated easily by its overall larger size. The side ridges on the upper surface of the pronotum are compressed or pinched together near the midpoint, but not as markedly as in *O. pelidna.* Also, the forewings are not spotted and speckled.

Distribution and Ecology. The handsome grasshopper is found throughout the state, although it appears more common in northern regions. It also is found widely in the United States west to the Rocky Mountains and south into Mexico. This species is commonly associated with dry grasses of short to moderate height. The males fly readily when disturbed; the females fly awkwardly and often escape by leaping. Both sexes stridulate.

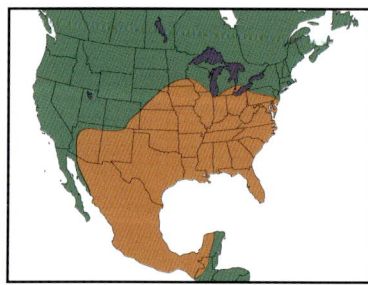

Map 10. Handsome grasshopper distribution.

Silent Slantfaced Grasshoppers

Subfamily ACRIDINAE

This is a very small subfamily in North America, with only one species known from the United States. However, several genera and numerous species occur in South America. Acridinae is very similar in appearance to the stridulating slantfaced grasshoppers (subfamily Gomphocerinae), but as the common name suggests, members of this subfamily lack stridulatory pegs on the hind femora of males and thus do not produce sound.

Grasshoppers in this subfamily have a slanted face, as is found in Gomphocerinae, and flattened, sword-shaped antennae, which also are present among some gomphocerines and spurthroated grasshoppers (subfamily Cyrtacanthacridinae). Acridines lack a spine between the front legs (the prosternal spine) that is found in the lubber grasshoppers (subfamily Romaleinae) and spurthroated grasshoppers. Also, the hind wings are colorless or nearly colorless, lacking the dark band found in the bandwinged grasshoppers.

Only one species occurs in Florida:

Metaleptea
> ***M. brevicornis* (Johannson)**

Metaleptea brevicornis
(Johannson)

Clippedwing
grasshopper

Plate 33. Clippedwing grasshopper (male).

Identification. This slantfaced species is distinguished by the angled tip of the forewings, a distinct contrast from the rounded or pointed wing tips found on nearly all other longwinged Florida species. The hind wings are transparent. Males of the species usually are green on top and brown on the sides. Females are more variable, often colored mostly brown or green, but sometimes light brown on the top and green on the sides. The antennae are

Plate 34. Clippedwing grasshopper (female).

markedly sword-shaped. The hind tibiae are brownish. Males measure 25–38 mm in length, females 36–53 mm.

Similar Species. This species is not readily confused with others. The bandwinged grasshopper *Dissosteira carolina* has wing tips that are similarly angled, but is not a slantfaced species, and bears black hind wings.

Distribution and Ecology. The clippedwing grasshopper has been collected throughout Florida, but though widespread, nowhere is it abundant. This species also occurs throughout eastern North America west to the Mississippi River, and in the South its distribution extends west into Texas. Its range includes most of Central and South America. The preferred habitat of clippedwing grasshopper is tall grasses along ponds and marshes. It also sometimes occurs in salt marshes. This species is a strong flier and, unlike most grasshoppers, sometimes is attracted to lights.

Map 11. Clippedwing grasshopper distribution.

Bandwinged Grasshoppers

Subfamily OEDIPODINAE

The bandwinged grasshoppers are usually heavybodied and bear enlarged hind legs. The head of these grasshoppers often appears enlarged and broadly rounded. The orientation of the face is nearly straight up and down, rather than slanting, in distinct contrast to species in the subfamily Gomphocerinae, the slantfaced grasshoppers. Bandwinged grasshoppers lack the spine between the front legs (the prosternal spine) that is found in the lubber grasshoppers (subfamily Romaleinae) and spurthroated grasshoppers (subfamily Cyrtacanthacridinae). The bandwinged grasshoppers tend to be gray or brown, and often are mottled with darker spots. The pronotum often bears ridges, wrinkles, or small tubercles or knobs, imparting a rough appearance.

The bandwinged grasshoppers usually bear bright colors, but this may not be obvious. The hind wings are often yellow, orange, or reddish where they join the body, with a broad black band crossing near the center of the wing. The colorful hind wings are hidden by the front wings (tegmina) except when in flight. Similarly, the inner face of the hind femora is often yellow, orange, red, or blue. Again, this is not usually apparent, and these species often blend exceptionally well with soil.

Males, and sometimes females, produce sound in flight (crepitation). The snapping, crackling, or buzzing sound is made by rubbing the undersurface of the forewings against the veins of the hind wings. They do not always crepitate in flight, as sound production plays a role in mate selection. Also, these grasshoppers sometimes produce sound while at rest (stridulation) by rubbing the hind femora against the forewings, but the femora lack the stridulatory pegs found in the stridulatory slantfaced grasshoppers (subfamily Gomphocerinae).

The oedipodine grasshoppers normally are associated with open, sunny areas, particularly areas with bare soil. Thin, overgrazed pastures or barren areas within pastures are the preferred habitat. They feed principally on grasses. When disturbed these grasshoppers fly readily, but alight on soil rather than on plants. Their general color often varies slightly, depending on the color of the soil in their environment. Thus, they can be very difficult to detect when they sit motionless on the ground.

There are 13 species in Florida, found in 8 genera:

Arphia
  A. granulata (Saussure)
  A. sulphurea (Fabricius)
  A. xanthoptera (Burmeister)
Chortophaga
  C. australior (Rehn and Hebard)
Dissosteira
  D. carolina (Linnaeus)
Hippiscus
  H. ocelote (Saussure)
Pardalophora
  P. phoenicoptera (Burmeister)
Psinidia
  P. fenestralis (Serville)
Spharagemon
  S. bolli Scudder
  S. crepitans (Saussure)
  S. cristatum (Scudder)
  S. marmorata (Scudder)
Trimerotropis
  T. maritima (Harris)

Arphia granulata
(Saussure)

Southern yellowwinged grasshopper

Plate 35. Southern yellowwinged grasshopper (male).

Identification. This is the common *Arphia* species in Florida. It is light to dark brown, often bearing small dark or black speckles on the forewings and elsewhere. The forewings usually bear a narrow, pale yellow hind margin on the forewings that forms a yellow stripe along the back when the wings are held at rest. In the field, the most distinctive feature of this grasshopper's appearance is the bright yellow hind wings. The hind wings also are marked with a curved black band. The hind tibiae are yellowish closer to the body, with a black band separating the basal third of the tibia from the second third. The two-thirds of the hind tibia farthest from the body is mostly pale or yellowish, but often contain some additional dark coloration. This portion of the tibiae is not usually mostly or entirely black, however. The ridge in the center of the upper surface of the pronotum is slightly, but distinctly, elevated. The length of males is 27–33 mm, whereas females measure about 30–35 mm.

Plate 36. Southern yellowwinged grasshopper with left wings spread.

Distribution and Ecology. *Arphia granulata* can be observed throughout most of the year in north Florida, and both adults and nymphs have been collected during the winter in south Florida. Presence of the adults is readily apparent because they make short, noisy flights in which they produce a crackling sound (crepitation) and flash their brightly colored wings. This grasshopper is found throughout Florida, and the southeastern states from Mississippi to North Carolina. It inhabits brushy fields, open woods, roadsides, and, to a lesser degree, grasslands.

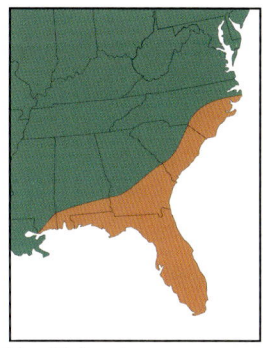

Map 12. Southern yellowwinged grasshopper distribution.

Arphia sulphurea (Fabricius)
Sulfurwinged grasshopper

Identification. This species greatly resembles *Arphia granulata,* but is rare in Florida. It is pale brown to dark brown, often bearing a narrow, pale yellow stripe on the hind margin of the forewings that forms a distinct stripe on the upper surface of the forewings when the insect is at rest. A sprinkling of darker spots often occurs, especially in the forewings. The hind wings are yellow near the body, with a curved black band crossing the wing, and the wing tip dusky, or dark. The hind tibiae are yellow closest to the body, with a black ring separating the basal third from the second third, and the region farther from the body variable. Adult males measure 23–31 mm in length, females 28–38 mm.

Similar Species. Sulfurwinged grasshopper is distinguished from the other two Florida *Arphia* species by the shape of the ridge (frontal costa) at the center of the grasshopper's face. In *A. sulphurea,* the ridge narrows markedly above the antennae, sometimes being as little as one-half the width of the ridge lower on the face. In contrast, *A. granulata* and *A. xanthoptera* have a facial ridge that is only slightly narrowed.

Map 13. Sulfurwinged grasshopper distribution.

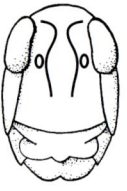

Figure 13. Face of *Arphia sulphurea (left)* and the nearly identical *A. granulata* and *A. zanthoptera (right).*

Distribution and Ecology. This species is widely distributed in North America east of the Great Plains. However, it apparently is rare in Florida, having been collected only in northern Florida, and only infrequently. The normal habitat is open pine woods containing scrub oaks. It overwinters in the nymphal stage, so adults are common in spring and early summer.

Plate 37. Autumn yellowwinged grasshopper (female).

Arphia xanthoptera
(Burmeister)

Autumn yellowwinged grasshopper

Identification. This is the largest of the *Arphia* species found in Florida, and its large size is a distinguishing characteristic. It has a brown to black-

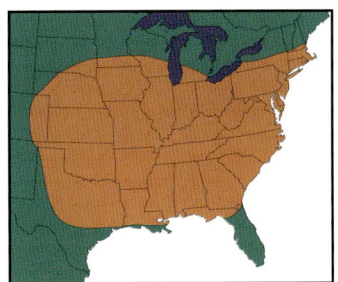

Map 14. Autumn yellowwinged
grasshopper distribution.

Plate 38. Autumn yellowwinged grasshopper
with left wings spread.

ish body with yellow (sometimes orange) hind wings that are marked with
a curved black band crossing the wing. The forewings are uniformly colored brownish to blackish. The ridge (frontal costa) at the center of the
face does not narrow markedly above the antennae. The ridge in the center
of the top of the pronotum is strongly elevated and arched. The tibiae are
mostly dark, including the region farthest from the body, with a pale ring
in the quarter closest to the body. Males of *A. xanthoptera* measure 31–38
mm, 36–46 mm in females.

Similar Species. The ridge in the middle of the pronotum of *A. xanthoptera*
is distinctly elevated and arched, and this serves as the most reliable diagnostic feature to separate it from other *Arphia* species. The dark distal
region—that portion farther from the body—of the tibiae is not entirely
consistent, but it is a fairly reliable character to separate *A. xanthoptera*
from *A. sulphurea,* and especially from *A. granulata.* The forewings of *A.
xanthoptera* lack the distinct yellowish stripe that is common on the other
Florida *Arphia.*

Distribution and Ecology. This species occurs in the autumn in northern
Florida, but is absent from the peninsula south of Orlando. It occurs
widely in the United States east to western Nebraska and Oklahoma. The
habitat of *A. xanthoptera* includes weedy borders of cultivated fields,
brushy fields, and open woods.

Plate 39. Southern greenstriped grasshopper (female).

Plate 40. Southern greenstriped grasshopper (female).

Chortophaga australior Rehn and Hebard

Southern greenstriped grasshopper

Identification. There are two color forms present in this species, a green form and a brown form, with intermediates—greenish brown or brownish green—found in both sexes. The principal difference between forms is found in the coloring of the head, thorax, and outer face of the hind femora. The ridge in the middle of the pronotum is slightly elevated. An X-shaped mark is present on the upper surface of the pronotum in the brown forms. The leading edge of the forewings is marked with 2–3 large green or light brown spots, with the balance of the forewings colored dark brown. The most important distinguishing character of this species is the color of the hind wing. Unlike Florida's other bandwinged species, southern greenstriped grasshopper lacks a bold, black band crossing the hind wing. The black band is present, but greatly muted, reduced to no more than a smoky area in many individuals. Similarly, the yellow in the the hind wing where it joins the body is muted to absent. The upper surface of the hind femora usually is marked with about 3 large dark spots; the central or largest spot is triangular when viewed from above. The hind tibiae are brown or bluish green. Males measure 21–25 mm in length, females 29–33 mm.

Distribution and Ecology. The range of the southern greenstriped grasshopper is Florida and adjacent southeastern states. It is found through-

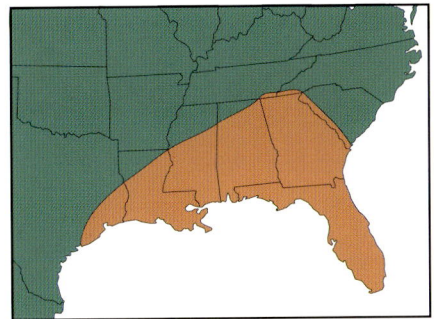

Plate 41. Southern greenstriped grasshopper with left wings spread.

Map 15. Southern greenstriped grasshopper distribution.

out Florida in open areas, but not in wooded sites. Favored habitats are old fields, heavily grazed pastures, and edges of crop fields and roadways.

Dissosteira carolina (Linnaeus)

Carolina grasshopper

Plate 42. Carolina grasshopper (female).

Identification. The color of Carolina grasshopper varies from yellowish gray to reddish brown, and it often bears numerous small dark spots over most of its body. A sharp ridge is found on the pronotum. The hind wings are black except for a marginal yellowish band and smoky gray wing tips. The black hind wings serve to distinguish this species from all other Florida grasshoppers. The hind tibiae are yellow. The males measure 37–42 mm in length, the females 42–48 mm.

Plate 43. Carolina grasshopper with left wings spread.

Distribution and Ecology. The Carolina grasshopper is a strong flier, and is often seen hovering or in the zigzag, fluttering flight of courtship.

It is easily mistaken for a butterfly when in flight. It is associated with barren soil such as dirt roadways and fallow fields. This species occurs in northern Florida and widely throughout the United States.

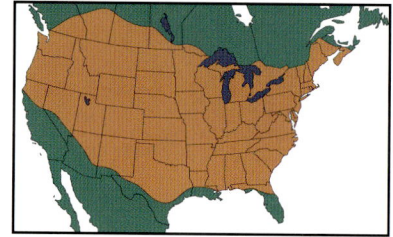

Map 16. Carolina grasshopper distribution.

Plate 44. Wrinkled grasshopper (male).

Hippiscus ocelote (Saussure)
Wrinkled grasshopper

Identification. This is a large, heavybodied species. It is gray and brown. The pronotum is usually rough or wrinkled, which is the basis of the common name. The top of the pronotum often has a light X-shaped mark, especially on males. The forewings bear large dark spots and light stripes; the latter come together at the tips to form a light-colored "V" on the upper surface when the wings are closed. The hind wings are usually

pale pinkish or orangish where they join the body, but sometimes tend toward yellow. The hind wings also have a broad dark band centrally, but with the tip poorly pigmented. The hind tibiae are yellow. Males measure 28–36 mm in length, females 39–44 mm.

Plate 45. Wrinkled grasshopper with left wings spread.

Distribution and Ecology. *Hippiscus ocelote* usually is found in pastures with thin or low-growing grass. It feeds on grass, and is an occasional pasture pest in Florida. Females are poor fliers,

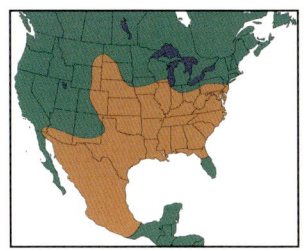

but males are active. This species is found in northern Florida, and occurs widely east of the Rocky Mountains.

Map 17. Wrinkled grasshopper distribution.

Plate 46. Orangewinged grasshopper (male).

Pardalophora phoenicoptera (Burmeister)

Orangewinged grasshopper

Identification. This is a large, gray and brown grasshopper with large dark spots on the forewings. The forewings also bear a light brown or gold diagonal stripe that forms a "V" when the grasshopper is viewed from above. Some individuals have green on the head, thorax, and hind femora. The

Plate 47. Orangewinged grasshopper (female).

basis of the common name is the bright orange or rose-colored hind wing, which also bears a broad, curved black band crossing centrally. The portion of the hind wing farthest from the body is smoky. The inner surface of the hind femora are bright blue and orange. The hind tibiae are orange. The males of this grasshopper measure 36–42 mm, females 45–55 mm.

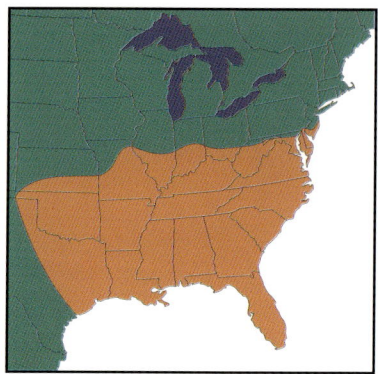

Plate 48. Orangewinged grasshopper with left wings spread.

Map 18. Orangewinged grasshopper distribution.

Distribution and Ecology. This species occurs throughout Florida and the eastern United States. It is obvious early in the season because the nymphs overwinter, and the adults are present in the spring, when few other grasshoppers are mature. Orangewinged grasshopper prefers an open habitat such as old fields and sandy areas, but may also be found in tall grass, brush, and wooded areas if plants are not too crowded. The male is an active flier, while the heavybodied female tends to remain on the soil. Sound production may occur on the ground (stridulation) or in flight (crepitation), but this is not a particularly noisy species.

Psinidia fenestralis (Serville)

Longhorn bandwinged grasshopper

Plate 49. Longhorn bandwinged grasshopper (female).

Identification. This small, thinbodied species is distinguished principally by its relatively long antennae. The antennal segments are somewhat flattened; where they join the body they are slightly larger. The general color

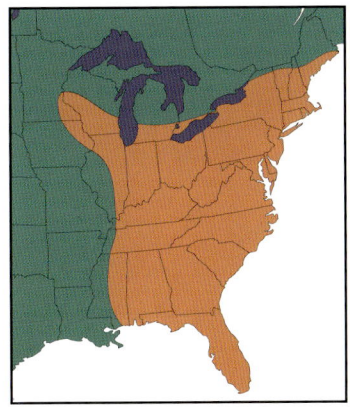

Plate 50. Longhorn bandwinged grasshopper with left wings spread.

Map 19. Longhorn bandwinged grasshopper distribution.

of *P. fenestralis* is usually gray and brown, but ranges from yellowish to blackish; its overall color tends to match its habitat. A narrow yellowish stripe runs from the back of the eye onto the pronotum. The leading edge of the forewings tends to be marked with alternating light and dark spots. The hind wings bear an unusually wide, curved black band centrally. The hind wings are usually orange closest to the body but sometimes are rose or yellow. The hind wing is variably smoky or partially blackened in the region farthest from the body. The hind tibiae are yellowish but bear a black band. The body length is 20–27 mm in males, 26–33 mm in females.

Distribution and Ecology. This species is found throughout Florida, and the United States east of the Mississippi River. Its habitat is open grassy areas, and specifically barren patches of sand within this general habitat. When disturbed, these grasshoppers fly only a short distance and alight on bare soil, where they blend in remarkably well with the background, becoming almost invisible. Males sometimes crepitate while flying. Adults or nymphs can be found throughout the year.

Plate 51. Boll's grasshopper (female).

Spharagemon bolli
Scudder

Boll's grasshopper

Identification. Boll's grasshopper is a grayish or reddish brown species, often covered with minute dark spots that blend together on the forewings to form scattered broad bands that cross the wings. The hind wings are crossed by a curved, black band centrally, are pale yellow closest to the body, and transparent or smoky away from the body. The ridge on top of the pronotum is slightly elevated. The outer face of the hind femora are weakly or indistinctly banded, but the inner face bears alternating black and pale yellow bands. The hind tibiae are yellowish closest to the body and reddish orange farther away, with a narrow black band separating the yellow and orange. The length of the males is 30–34 mm, and females measure 35–45 mm.

Similar Species. The narrowness of the black band on the hind tibiae is useful in separating this species from the similar *Spharagemon crepitans,* which has a broader black tibial band.

Plate 52. Boll's grasshopper with left wings spread.

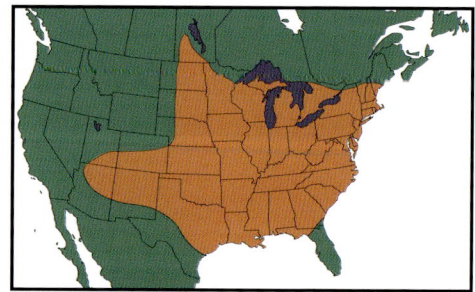

Map 20. Boll's grasshopper distribution.

Distribution and Ecology. Although this species is found widely in the United States, from the Atlantic Ocean to the Rocky Mountains, in Florida it is known only from the northwestern Panhandle region. The preferred habitat is open, sunny woods, although sometimes it is found along the margins of woods. Males crepitate and stridulate frequently, and often hover about one meter above the ground while displaying.

Spharagemon crepitans (Saussure)

Crepitating grasshopper

Plate 53. Crepitating grasshopper (female).

Identification. This grayish brown or reddish brown grasshopper occasionally has scattered broad dark bands across the forewings, but usually they are lacking. The hind wings are pale yellow at the base, but are crossed by a wide, curved black band. Farther from the body, the hind wing is smoky or colorless. The ridge in the middle of the pronotum is slightly elevated. The hind tibiae are yellowish at their bases and reddish orange farther out, with a broad black band centrally. The length of males is 30–34 mm, females 37–42 mm.

Plate 54. Crepitating grasshopper with left wings spread.

Similar Species. The black band across the hind wing is wider, and located more centrally, than the corresponding band in *Spharagemon bolli*. The width of the black band on the hind tibiae is similar to the width of the orange distal portion, considerably wider than in *S. bolli*, a very similar species.

Distribution and Ecology. The crepitating grasshopper is confined almost entirely to Florida, although a few specimens have been collected from southern Georgia. It occurs widely in the state and has been collected from diverse habitats. Surprisingly, it may be found in oak woods in shaded areas not typically inhabited by grasshoppers.

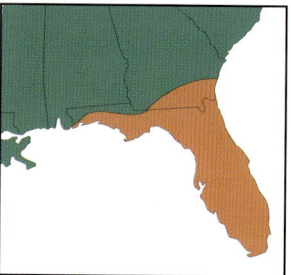

Map 21. Crepitating grasshopper distribution.

Plate 55. Ridgeback sand grasshopper.

Spharagemon cristatum (Scudder)

Ridgeback sand grasshopper

Identification. This grayish or brownish grasshopper is heavily mottled with black spots. The speckling on the forewings can be organized into irregular bands that cross the wings and into black tips. The sharp ridge found on top of the pronotum is higher than on any other Florida species. The hind wings of ridgeback sand grasshopper are yellow where they join the body, with a curved black band centrally, and a colorless or smoky wing tip. The hind femora are speckled brown on the outer face, but the inner face bears alternating bands of black and yellow. The hind tibiae are pale yellow at their bases, but principally dark orange or red. The body length is 29–39 mm in males, 34–45 mm in females.

Similar Species. The only other species that have nearly such an elevated pronotal ridge are *Dissosteira carolina,* Carolina grasshopper, and *Arphia xanthoptera,* autumn yellowwinged grasshopper. However, *D. carolina* has black hind wings, and *A. xanthoptera* is uniformly dark, so they are readily distinguished.

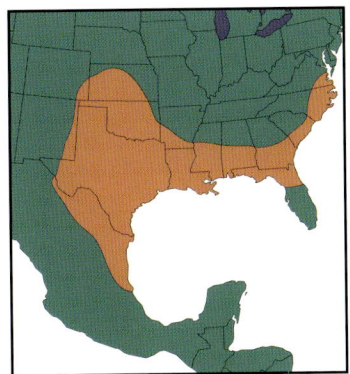

Plate 56. Ridgeback sand grasshopper with left wings spread.

Map 22. Ridgeback sand grasshopper distribution.

Distribution and Ecology. The ridgeback sand grasshopper is found throughout northern Florida south to about Orlando. It also occurs in adjacent southeastern states and in the southern Great Plains. The common habitat is unused crop fields, old fields, margins of woods, and sandy roadsides.

Spharagemon marmorata Scudder)

Marbled grasshopper

Plate 57. Marbled grasshopper (female).

Identification. The marbled grasshopper is gray and brown, but has well-marked, blackish bands on the leading edge of the forewings that cross the wings and merge into a solid black trailing edge. The hind wings of *S. marmorata* bear an unusually wide, curved black band centrally. The base of the hind wings is orange-yellow or dark yellow, and farther out they are smoky or partially blackened. The hind tibiae are orange or red, with

Plate 58. Marbled grasshopper with left wings spread.

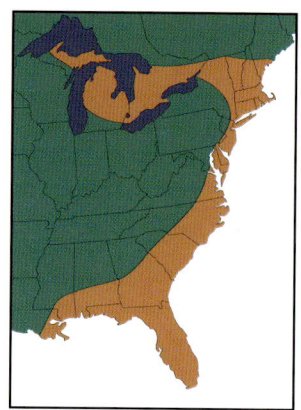

Map 23. Marbled grasshopper distribution.

a pale yellowish ring at the base. The body length of marbled grasshopper is 26–31 mm in males, and 27–35 mm in females.

Similar Species. The banding pattern of the forewings helps distinguish this species from *Psinidia fenestralis,* a co-occurring species in sandy habitats. Tibia color also serves to distinguish *S. marmorata* from *P. fenestralis,* the latter species having yellow and black hind tibiae.

Distribution and Ecology. The marbled grasshopper can be found throughout Florida. It also occurs in adjacent states, along the east coast to New England, and in the United States and Canada surrounding the Great Lakes. This species frequents open, sandy areas. Sand dunes along beaches, disturbed areas of pastures, and sunny, sandy areas of open woods commonly are inhabited by this species. Males crepitate loudly during their lengthy, zigzag flights. While on the ground they also stridulate and make complicated leg-lifting movements as part of their courtship ritual. These grasshoppers can often be found throughout the year in Florida.

Trimerotropis maritima (Harris)

Seaside grasshopper

Identification. This grasshopper is light gray to dark grayish brown. It bears numerous small brown speckles over most of its body, and weakly to strongly marked wide bands across the forewings. The hind wings are pale yellow where they join the body, and marked with a curved black band

Plate 59. Seaside grasshopper (male).

through the center. The distal portion of the hind wing is transparent. The ridge in the middle of the pronotum is barely elevated. The outer face of the hind femora is gray and brown, with only weak evidence of bands. The inner face, however, is pale yellow with three black bands. The hind tibiae are yellow to red. The length of this species is 29–33 mm in males and 30–40 in females.

Distribution and Ecology. Found throughout Florida, this species also occurs throughout the eastern United States west to the Rocky Mountains. This is a sand-loving grasshopper, found in arid, barren areas. A common resident of ocean, lake, and river margins, *T. maritima* also frequents unused crop fields and sandy roadways. When disturbed it is likely to crouch motionless, blending well with its sandy background. It is a strong flier, however, and can travel long distances. Males display both crepitation and stridulation in their courtship ritual.

Plate 60. Seaside grasshopper with left wings spread.

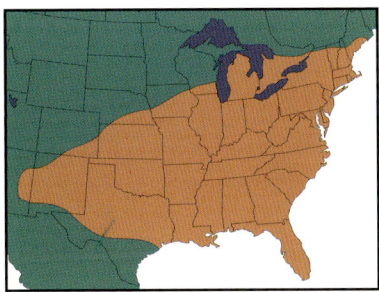

Map 24. Seaside grasshopper distribution.

Spurthroated Grasshoppers

Subfamily CYRTACANTHACRIDINAE

This subfamily is sometimes divided, with members of the genus *Schisto-cerca* placed in the subfamily Cyrtacanthacridinae, and several other related genera comprising the subfamily Melanoplinae. Here we do not recognize this division and retain the traditional, broader classification system.

The spurthroated grasshoppers, as the common name suggests, bear a spur or spine (the prosternal spine) between the front legs. The antennae usually are threadlike, not flattened or sword-shaped. The head is not especially large, and these grasshoppers do not appear to be especially heavybodied. In most genera, the head is oriented straight up and down, but in some groups the face is slanted. These grasshoppers may be wingless, or may bear short or long wings. The forewings (tegmina) are pigmented, but lack the bands crossing them that are common in the bandwinged grasshoppers, subfamily Oedipodinae. The hind wings are transparent. The flying ability of spurthroated grasshoppers varies greatly, even within a single genus. The genus *Schistocerca* contains especially long-winged, strong fliers. They are sometimes called bird grasshoppers in recognition of their large size and strong flying abilities.

The subfamily Cyrtacanthacridinae is large and diverse, and several species are known only from Florida. The largest group is the genus *Melanoplus,* with more than 200 species in North America and 26 species in Florida. It is sometimes difficult to distinguish among *Melanoplus* species, and to a lesser degree among *Schistocerca* species. Thus, the sexual or terminal abdominal structures of the males are commonly used to distinguish among similar-appearing species. The characters of interest are the paired cerci; the forked, two-lobed furcula; the supra-anal plate; and the subgenital plate.

The habitat of these grasshoppers is highly variable. Although some are found in trees or in the undergrowth of dense woods, most are found in open grassy or weedy areas. The dietary habits vary from monophagy, or feeding strictly on one plant species or genus, to polyphagy, or feeding on plants from several plant families. Species may specialize on grasses, forbs, shrubs, or trees, or may feed freely among all these plant types.

There are 45 species of Cyrtacanthacridinae in Florida, found in 9 genera:

Aptenopedes
  A. aptera Scudder
  A. sphenarioides Scudder
Eotettix
  E. palustris Morse
  E. pusillus Morse
  E. signatus Scudder
Gymnoscirtetes
  G. morsei Hebard
  G. pusillus Scudder
Hesperotettix
  H. floridensis Morse
  H. osceola Hebard
  H. viridis (Thomas)
Leptysma
  L. marginicollis (Serville)
Melanoplus
  M. adelogyrus Hubbell
  M. apalachicolae Hubbell
  M. bispinosus Scudder
  M. davisi (Hebard)
  M. forcipatus Hubbell
  M. furcatus Scudder
  M. gurneyi Strohecker
  M. impudicus Scudder
  M. indicifer Hubbell
  M. keeleri (Thomas)
  M. nanciae Deyrup
  M. ordwayae Deyrup
  M. propinquus Scudder
  M. puer (Scudder)
  M. punctulatus Scudder
  M. pygmaeus Davis
  M. querneus Rehn and Hebard
  M. rotundipennis Scudder

M. sanguinipes (Fabricius)

M. scapularis Rehn and Hebard

M. scudderi (Uhler)

M. strumosus Morse

M. symmetricus Morse

M. tepidus Morse

M. tequestae Hubbell

M. withlacoocheensis Squitier and Deyrup

Paroxya

P. atlantica Scudder

P. clavuliger (Serville)

Schistocerca

S. alutacea (Harris)

S. americana (Drury)

S. ceratiola Hubbell and Walker

S. damnifica (Saussure)

S. obscura (Fabricius)

Stenacris

S. vitreipennis (Marschall)

Aptenopedes aptera Scudder

Wingless Florida grasshopper

Plate 61. Wingless Florida grasshopper (male).

Identification. This green grasshopper has a distinctly slanted face. The presence of a spine or spur underneath, between the front legs, separates it from slantfaced grasshoppers, with which it is easily confused. It is also easy to confuse the nymphal and adult stages, because the adult shows virtually no evidence of wings. Obviously this species is flightless. This grasshopper usually is well marked with a yellowish or reddish stripe on the side of the pronotum, and the males usually have a distinct white stripe on the top of the abdomen. The hind tibiae are bluish green. Females are much larger and more heavybodied than males. The males measure 15–21 mm in length, the females 21–33 mm.

Similar Species. *Hesperotettix oceola* bears a strong resemblance to *A. aptera* because the general body form and color patterns are similar. However, the short wings of adult *H. osceola* should serve to distinguish it from wingless Florida grasshopper.

Plate 62. Wingless Florida grasshopper (male).

Distribution and Ecology. The wingless Florida grasshopper is found throughout Florida in very dry areas, particularly in association with oak shoots and dwarf oaks in scrub habitats. It also is found in adjacent southeastern states.

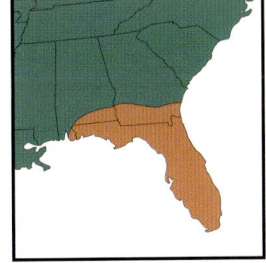

Map 25. Wingless Florida grasshopper distribution.

Aptenopedes sphenarioides Scudder

Linearwinged grasshopper

Plate 63. Linearwinged grasshopper (female).

Identification. This species is usually green, but sometimes tends toward purplish brown. Like *Aptenopedes aptera,* this species has a strongly slanted face, and superficially resembles slantfaced grasshoppers, subfamilies Gomphocerinae and Acridinae. It bears the spine or spur between the front legs, however, that marks the subfamily Cyrtacanthacridinae. This grasshopper, in the adult stage, has its wings reduced to long pads along the

Plate 64. Linearwinged grasshopper (male).

sides, which is the basis of the common name. This species is flightless. A light and dark stripe is usually present on the sides, running from the top of the eye to the tip of the wing pads. A similar stripe may occur down the center of the back, though this is much more frequent among males than females. The hind tibiae are bluish green. Females are considerably larger and more heavybodied than males. The males measure 16–23 mm in length, the females 22–30 mm.

Similar Species. *Hesperotettix osceola* bears a strong resemblance to *A.*

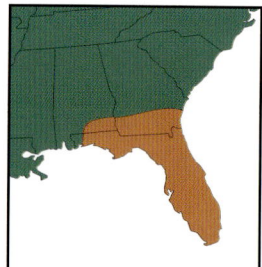

Map 26. Linearwinged grasshopper distribution.

sphenarioides because the general body form and color patterns are similar. However, the short but obvious wings of adult *H. osceola* should serve to distinguish it from the linearwinged grasshopper, which bears only a small flap or pad rather than a wing.

Distribution and Ecology. This is a fairly common grasshopper found throughout Florida among grass and shrubs in open areas and open high pine, flatwood, and hammock habitats. It also occurs in adjacent southeastern states.

Eotettix palustris Morse

Swamp eastern grasshopper

Identification. This small but attractive grasshopper is metallic yellowish green or brownish green. A broad black stripe connects the eye to the rear edge of the pronotum. The hind femora are yellowish, the hind tibiae pale reddish. The forewings of this flightless grasshopper are oval and shorter than the prothorax. The length of the body is 15–15.5 mm in males and 21–22 mm in females.

Similar Species. The oval forewings serve to distinguish this species from the similar *Eotettix pusillus,* which has forewings that are nearly round. It can also be distinguished from *E. signatus,* which has forewings that are longer than the pronotum and which taper to a rounded but distinct point.

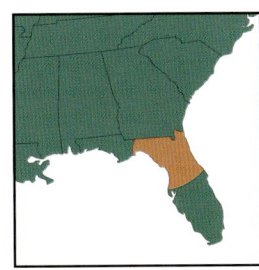

Map 27. Swamp eastern grasshopper distribution.

Distribution and Ecology. The distribution of the swamp eastern grasshopper apparently is restricted to northern Florida. It is found in moist areas of pine woods.

Eotettix pusillus Morse

Little eastern grasshopper

Plate 65. Little eastern grasshopper (female).

Identification. This metallic grasshopper is strikingly small, males measuring only 10–15 mm in length and females 15.5–20 mm. It is metallic yellowish green, brownish green, or reddish brown. The pronotum bears a black spot on each side behind the eye, but it does not extend completely to the rear edge of the pronotum. The hind femora and tibiae are reddish gold or greenish gold. Most abdominal segments are partially black, which results in a black vertical banding pattern. The forewings of this flightless species are almost round, an important distinguishing character, and shorter than the pronotum. Interestingly, the nymphal stage differs completely in color, possessing a black body with a red and gold head.

Similar Species. The black spot on the side of the pronotum does not extend to the hind margin of the pronotum, as occurs in *Eotettix palustris*

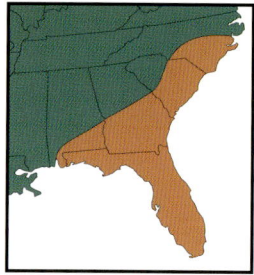

Map 28. Little eastern grasshopper distribution.

and *E. signatus.* The hind tibiae of the other Florida *Eotettix* species are reddish, whereas in *E. pusillus* they tend toward gold.

Distribution and Ecology. The little eastern grasshopper occurs in northern Florida and other southeastern states. It is found in open pine and oak woods.

Plate 66. Handsome Florida grasshopper (female).

Eotettix signatus Scudder

Handsome Florida grasshopper

Identification. This species is metallic yellowish green or bluish green. The forewings of this flightless species are elongate oval and as long as, or longer than, the pronotum. This species is marked on each side with a black stripe running from the eye to the rear edge of the pronotum. The hind tibiae are red. The males measure 18–21 mm in length, the females 19.5–30.6 mm.

Similar Species. The length and shape of the forewings serve to distinguish *E. signatus* from *E. palustris* and *E. pusillus.* The latter two species have

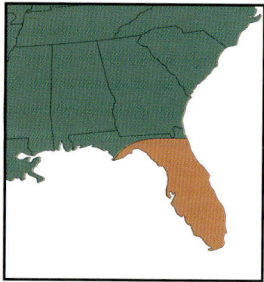

Map 29. Handsome Florida grasshopper distribution.

round to oval forewings that are shorter than the pronotum.

Distribution and Ecology. *Eotettix signatus* occurs throughout Florida, but its distribution apparently is restricted to Florida. The habitat of this species is variable. It can be found in moist or fairly dry areas, in open pine forests or on prairie. Sometimes it is abundant adjacent to ponds.

Gymnoscirtetes morsei Hebard

Morse's wingless grasshopper

Identification. This species is wingless, greenish yellow to tan, and bears a blackish lateral stripe running from each eye to about the midpoint of the abdomen. Thus it is very similar in appearance to *Gymnoscirtetes pusillus* Scudder. The male measures 14–16 mm in length, the female 19.5–21.5 mm.

In males, the cerci taper gradually to a sharp point, but the top edge is strongly curved. The furcula is apparent, but not elongate. The tip of the subgenital plate is extended and elevated. It is about twice as high as it is wide. The side edges of the subgenital plate also are strongly elevated.

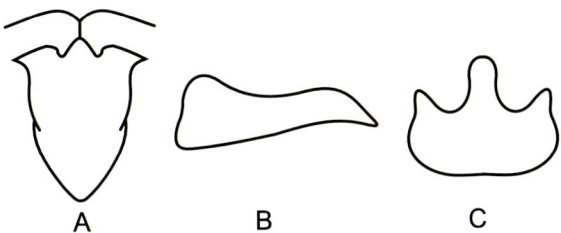

A   B   C

Figure 14. Supra-anal plate and furcula *(A)*, cercus *(B)*, and subgenital plate *(C)* of male *G. morsei.*

Similar Species. *Gymnoscirtetes morsei* is slightly larger than *G. pusillus*. In *G. pusillus* the tubercle at the tip of the subgenital plate in males is about as broad as high, whereas in *G. morsei* the tubercle is twice as high as wide. The side edges of the subgenital plate are strongly elevated, a character absent from *G. pusillus*. In addition, the top edge of the cerci in *G. pusillus* is relatively straight, orienting downward only slightly at the tip, whereas in *G. morsei* the top edge is strongly curved.

Distribution and Ecology. Morse's wingless grasshopper is known only from northern Florida and southern Alabama, but may inhabit nearby southern Georgia as well. It is found in wet areas and among wire grass in pine forests.

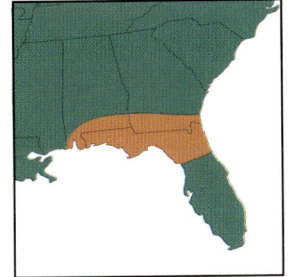

Map 30. Morse's wingless grasshopper distribution.

Plate 67. Little wingless grasshopper (female).

Gymnoscirtetes pusillus
Scudder

Little wingless grasshopper

Identification. This small grasshopper is greenish yellow to tan in color, and bears a black and white stripe along the sides from each eye to about the midpoint of the abdomen or beyond. The adult is wingless. Little wingless grasshopper greatly resembles *Gymnoscirtetes morsei* Hebard, Morse's wingless grasshopper. The males measure 12.5–15 mm in length and the females 17–22 mm.

In males, the furcula is barely visible. The cerci taper gradually to a point, but the top edge is almost straight. The tip of the subgenital plate is slightly extended into a tubercle about as high as it is wide.

Similar Species. In *G. pusillus* the tubercle at the tip of the subgenital plate in males is about as broad as high, whereas in *G. morsei* the tubercle is twice as high as wide. The side edges of the subgenital plate are not elevated, as occurs in *G. morsei*. Also, the cercus of *G. pusillus* is relatively straight, with only the tip curved slightly downward, whereas in *G. morsei* it is strongly curved.

Distribution and Ecology. The habitat of the little wingless grasshopper is wet areas of pine forests, where it inhabits the undergrowth, or adjacent to ponds. This is an agile species that easily eludes capture. It has been found throughout Florida except for the southernmost areas, and is also known in Georgia.

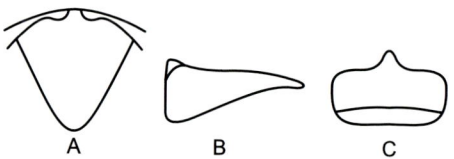

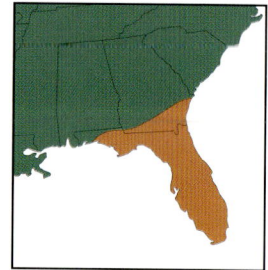

Figure 15. Supra-anal plate and furcula *(A)*, cercus *(B)*, and subgenital plate *(C)* of male *G. pusillus*.

Map 31. Little wingless grasshopper distribution.

Hesperotettix floridensis
Morse

Florida purplestriped grasshopper

Plate 68. Florida purplestriped grasshopper (female).

Identification. This is a fairly large, heavybodied grasshopper. Despite the common name, it is principally bright green. The only purple stripe is a narrow stripe on top of the pronotum, and this is often absent. Portions of the femora tend to be purplish, however. The pronotum is a large structure, and rather rough in texture. The forewings are oval and short, measuring about one and a half times as long as broad. The hind tibiae are green or bluish green. Males of this species measure 17.5–21 mm, females 24–30 mm.

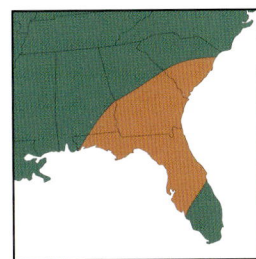

Map 32. Florida purplestriped grasshopper distribution.

Distribution and Ecology. Florida purplestriped grasshopper is apparently restricted to northern Florida and adjacent areas of Georgia. Its habitat is brushy and wet areas, and open woods.

Plate 69. Osceola's grasshopper (female).

Hesperotettix osceola
Hebard

Osceola's grasshopper

Identification. This is a principally green species, but the forewings are colored by a narrow stripe of reddish purple edged with white. On the top, a white stripe bordered by dark stripes extends the length of the pronotum and abdomen, and a short black-and-white stripe occurs on the sides behind each eye. The forewings are usually shorter than the pronotum. The hind tibiae are greenish blue. Males measure 14–19 mm in length, females 18–21 mm.

Map 33. Osceola's grasshopper distribution.

Similar Species. The short forewings and white abdominal stripe distinguish this species from the similar *H. viridis. Hesperotettix osceola* also resembles *Aptenopedes aptera* and *A. sphenarioides,* but the *Aptenopedes* species lack true wings.

Distribution and Ecology. This uncommon species is found only in Florida. Its habitat is scrub oak woods and other open, dry locations.

Plate 70. Meadow purplestriped grasshopper (female).

Hesperotettix viridis (Thomas)

Meadow purplestriped grasshopper

Identification. This is a colorful species, although the eastern form found in Florida is much less striking than the form found in western states. It is principally green, but the forewings are colored by a broad stripe of reddish purple edged with white. Purplish coloration is found along the top of the abdomen and sometimes occurs on the sides of the pronotum of some individuals. A short black-and-white stripe occurs on the sides behind each eye. The light stripe on the upper surface of the pronotum is bordered by dark stripes. The forewings may be long, reaching the tip of the abdomen or slightly beyond, or may be shorter, extending about two-thirds the length of the abdomen. The hind tibiae are blue. The length of males is 16–18 mm; females measure 18–28 mm.

Similar Species. The forewings are always longer than the pronotum in this species, a character that distinguishes it from *H. osceola*. *Hesperotettix viridis* is quite variable in appearance, however, so its specific status is subject to debate, and it has acquired several names.

Distribution and Ecology. The meadow purplestriped grasshopper is found in northern Florida, and throughout most of the United States. It normally is found inhabiting weedy and brushy locations, particularly dry habitats.

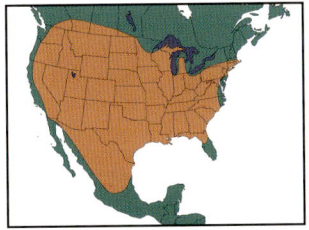

Map 34. Meadow purplestriped grasshopper distribution.

Plate 71. Cattail toothpick grasshopper (female).

Leptysma marginicollis (Serville)

Cattail toothpick grasshopper

Identification. This slender, elongate grasshopper has a very pointed head and flattened, sword-shaped antennae. Thus, it superficially resembles grasshoppers in the subfamily Gomphocerinae, but is easily distinguished by the presence of the spur or spine between the front legs. Cattail toothpick grasshopper is usually brownish with a whitish, yellow, or brown stripe from the eye to the base of the front legs. The head is as long as, or longer than, the pronotum. On top, the body may also be reddish or pinkish. The front wings are sharply pointed, extending 3–5 mm beyond the tip of the abdomen. The body length is 28–31 mm in males and 31–38 mm in females.

Similar Species. This species is easily confused with *Stenacris vitreipennis,* but in *L. marginicollis* the head is at least as long as the pronotum, whereas in *S. vitreipennis* the head is shorter than the pronotum. The antennal segments are considerably wider than in *S. vitreipennis.*

Distribution and Ecology. The cattail toothpick grasshopper inhabits wet areas, and is usually found on emergent vegetation such as cattails

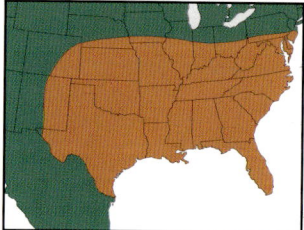

and sedges. Adults fly readily when disturbed but never alight on soil. They usually land on the stem of emergent vegetation and move quickly to the side opposite the source of disturbance. This species is known throughout Florida except for the Keys, and also occurs in other eastern and southwestern states.

Map 35. Cattail toothpick grasshopper distribution.

Melanoplus adelogyrus Hubbell
St. Johns spurthroat grasshopper

Identification. This is a small, rare short-winged species. The forewings are shorter than the pronotum, and elongate-oval. The forewings are widely separated on top. The general color is reddish brown or purplish gray above, and yellowish below. The males have behind the eye a shiny black stripe that extends across the pronotum to the first abdominal segments, but this pattern is indistinct in females. This black stripe is very wide on the front portion of the pronotum, narrowing markedly on the rear region of the pronotum. The hind femora are dull yellow, sometimes with black spots but not complete bands. The hind tibiae are purplish green. The males measure 12.5–15 mm, the females 17–21 mm.

In males, the furcula is very short. The top edge of the cerci is depressed toward the midpoint of the cerci, with the tip expanded slightly, bluntly rounded, and flattened. The subgenital plate is only weakly elongated.

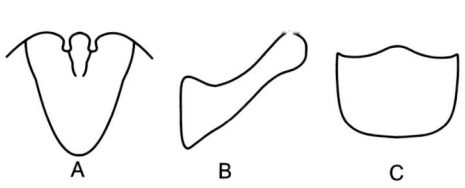

Figure 16. Supra-anal plate and furcula *(A)*, cercus *(B)*, and subgenital plate *(C)* of male *M. adelogyrus*.

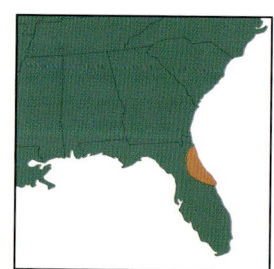

Map 36. St. Johns spurthroat grasshopper distribution.

Similar Species. The shape of the black stripe on the pronotum is important in distinguishing *Melanoplus adelogyrus* from *M. apalachicolae, M. gurneyi, M. ordwayae,* and *M. tequestae.* The flattened tip of the cerci is a useful character for separating *M. adelogyrus* from the very similar *M. puer.*

Distribution and Ecology. *Melanoplus adelogyrus* is found only in high pine and scrub habitat in northeastern Florida, east of the St. Johns River in Volusia County. Apparently this broad river has served as an ecological barrier, allowing *M. adelogyrus* to evolve and differentiate from *M. puer.* This latter species is found west of the St. Johns River.

Melanoplus apalachicolae Hubbell

Apalachicola spurthroat grasshopper

Plate 72. Apalachicola spurthroat grasshopper (male).

Identification. A small short-winged species, *Melanoplus apalachicolae* is closely related to *M. puer,* and very similar in appearance to *M. tequestae.* The forewings are shorter than the pronotum, elongate-oval, and widely separated on top of the body. The general color is reddish brown or purplish gray above, and yellowish below. The pronotum is more elongate and narrow than in related species. The males have a shiny black stripe behind the eye that extends across the pronotum, but this pattern is indistinct in females. This black stripe is narrow and has parallel sides on the front portion of the pronotum, widening slightly on the rear region of the pronotum. The hind femora are dull yellow, sometimes with black spots but not complete bands. The hind tibiae are purplish green. The males measure 13–14.5 mm in length, the females 17–20.5 mm.

In males, the furcula is short and rounded, but well developed. The cerci are tapered, but although the lower edges taper evenly throughout their length, the upper edges are tapered abruptly at their bases, leaving the remainder of the top edges almost straight.

Similar Species. The shape of the black stripe on the side lobe of the pronotum is an important character to distinguish *M. apalachicolae, M. gurneyi, M. ordwayae,* and *M. tequestae* from *M. puer* and *M. adelogyrus.* The presence of a furcula serves to distinguish *M. apalachicolae* from *M. tequestae* and *M. ordwayae,* which lack a visible furcula. The abrupt tapering of the upper edge at the base of the cerci distinguish *M. apalachicolae* from *M. gurneyi,* in which both the upper and lower margins taper gradually and symmetrically to the tip.

Distribution and Ecology. This species is found in the sandy uplands west of Tallahassee, in Gadsden and Liberty counties in northwest Florida. The high pine habitat is mostly turkey oak, blue jack oak, or longleaf pine, with an understory of wire grass and oak seedlings.

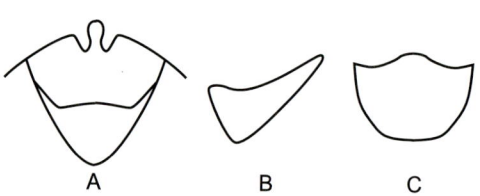

Figure 17. Supra-anal plate and furcula *(A)*, cercus *(B)*, and subgenital plate *(C)* of male *M. apalachicolae.*

Map 37. Apalachicola spurthroat grasshopper distribution.

Plate 73. Twospined spurthroat grasshopper (male).

Melanoplus bispinosus
Scudder

Twospined spurthroat grasshopper

Identification. This medium-sized *Melanoplus* is grayish brown to reddish brown. A dark bar extends from the eye onto the side lobe of the pronotum. The front wings are marked with a row of dark spots centrally. The forewings extend to the tip of the abdomen or beyond. The hind femora bear large dark spots that do not completely form stripes. The hind tibiae are bluish green or blue. Males measure 25–30 mm in length, females 26–32 mm.

In males, the furcula is V-shaped, and extends to about one-half the length of the supra-anal plate. The large "spinelike" furcula apparently is the basis for the name of this grasshopper. The cerci are elongate, narrowed at the middle and rounded at the ends; the outer face of the tip is grooved or recessed.

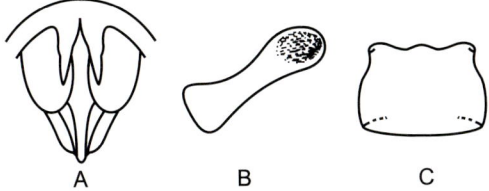

A   B   C

Figure 18. Supra-anal plate and furcula (A), cercus (B), and subgenital plate (C) of male *M. bispinosus*.

Distribution and Ecology. This species is found through most of Florida, though rarely in large numbers. It occurs in other southeastern states and west to Texas and Oklahoma. Twospined spurthroat grasshopper is found in pastures, crop fields, and roadsides.

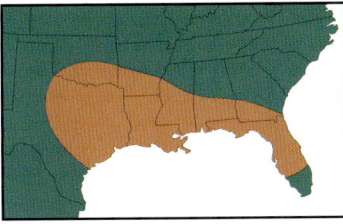

Map 38. Twospined spurthroat grasshopper distribution.

Melanoplus davisi (Hebard)

Davis' oak grasshopper

Plate 74. Davis' oak grasshopper (male).

Identification. This is among the largest of the short-winged *Melanoplus*, but otherwise indistinct. General color is brown, olive, or yellowish brown above and yellowish below, with a black stripe extending from the eye onto the pronotum. The forewings are oval, shorter than the pronotum, and overlap on top of the body. The hind femora are reddish yellow, with two black bars across them that tend to be present in males but absent in females. The hind tibiae are red. The males measure 18–22 mm, the females 25–27 mm.

In males, the furcula consists only of minute, rounded appendages. The cerci are broad and short, turning upward at the end to a flattened, blunt tip.

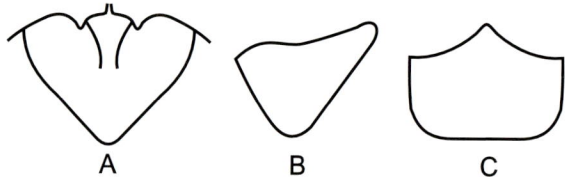

Figure 19. Supra-anal plate and furcula (A), cercus (B), and subgenital plate (C) of male *M. davisi*.

Similar Species. *Melanoplus davisi* shares the overlapping forewing character with *M. scapularis, M. strumosus,* and *M. scudderi.* It is difficult to confuse *M. davisi* with *M. scapularis* due to the shape of the cerci, which are expanded at the ends in the latter species. Similarly, in *M. strumosus* the cerci are long, slender, and constricted at the middle, so this species is easily distinguished. *Melanoplus scudderi* is more similar, with both species having cerci that turn upward at their ends into blunt tips. However, the furcula of *M. davisi* appear to be minute rounded appendages, whereas in *M. scudderi* they are pointed. Also, the tips of the cerci are much wider in *M. scudderi* than in *M. davisi*.

Distribution and Ecology. This grasshopper is found in northern Florida, but apparently does not occur in adjacent states. High pine is the favored habitat, where it feeds on understory, particularly low-growing oak.

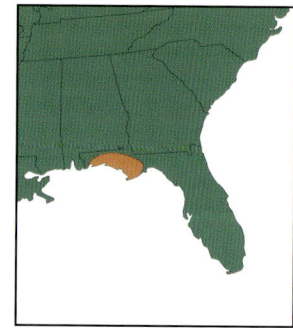

Map 39. Davis' oak grasshopper distribution.

Melanoplus forcipatus Hubbell

Toothcercus spurthroat grasshopper

Plate 75. Toothcercus spurthroat grasshopper (male).

Identification. This small short-winged species is very similar in appearance to *M. nanciae* and related species in the *M. puer* group. The forewings are shorter than the pronotum, oval, and widely separated above. The general color is reddish brown or purplish gray above, and yellowish below. The males have behind the eye a shiny black stripe that extends across the pronotum and is nearly equal in width from the front to the rear edge of the side lobe. This pattern is indistinct in females. The hind femora are dull yellow, sometimes with black spots but not complete bands. The hind tibiae are purplish green. The males measure 11.5–14.5 mm in length, the females 17–20.5 mm.

In males, the furcula is not visible. The cerci are distinctive and serve to identify this species readily. At their bases the cerci are broad, with the upper and lower margins nearly parallel. Beyond the middle, however, the cerci fork into two projections: short and stubby above, and long, tapering, downward-curving, and flattened below. The cerci also curve inward markedly and bear small teeth or projections on the upper surface or on the inner face. The supra-anal plate is less broad and less shieldlike than in many species, with the plate tapering only slightly near the furcula.

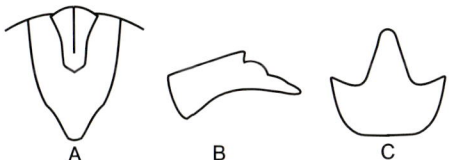

Figure 20. Supra-anal plate and furcula *(A)*, cercus *(B)*, and subgenital plate *(C)* of male *M. forcipatus*.

Similar Species. *Melanoplus forcipatus* closely resembles *M. indicifer.* However, the cercus of *M. forcipatus* bears a relatively prominent, broad, and strongly curved projection underneath, and small toothlike projections on the upper side and on the inner face. In contrast, the cercus of *M. indicifer* is less prominent, not usually bearing teeth on the top and on the inner face, and with the projection underneath narrow and not strongly curved. Although these two species are closely related, they do not occur in the same geographic areas of Florida, with *M. indicifer* restricted to the east coast of Florida near Palm Beach.

Distribution and Ecology. The distribution and habitat of *Melanoplus forcipatus* is largely the same as that of *M. tequestae,* the scrub oak habitat of the sandy ridges of central Florida.

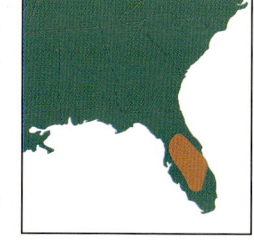

Map 40. Toothcercus spurthroat grasshopper distribution.

Melanoplus furcatus Scudder
Larger forktailed grasshopper

Identification. This is a large, heavybodied, longwinged species. It is generally brownish, but frequently with reddish or dark brown, especially on top. The dark bar commonly found behind the eye of *Melanoplus* species, extending along the side of the pronotum, may be present or absent. The forewing is darker near the body and lighter farther out. Small dark spots may be present or absent centrally on the forewing. The hind femora may bear a dark band. The hind tibiae are dull red. The males measure about 31 mm in length, the females about 39 mm.

In males, the furcula is not visible. The cerci are stout and forked, providing the basis for the common name of this grasshopper. The cerci taper from the base to about the midpoint, then fork into upper and lower, bluntly pointed triangular structures.

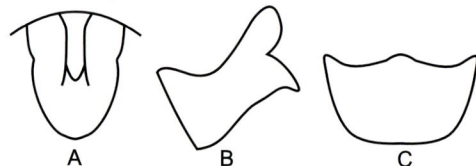

Figure 21. Supra-anal plate and furcula *(A),* cercus *(B),* and subgenital plate *(C)* of male *M. furcatus.*

Similar Species. *Melanoplus furcatus* is closely related to *M. symmetricus,* and may eventually be shown to be a form of that species. Presently they are distinguished by the shape of the cercus; in *M. symmetricus,* the cercus is symmetrical, expanding uniformly upward and downward. In *M. furcatus,* the expansion is V-shaped and not symmetrical.

Distribution and Ecology. This species is found in northeast Florida and southern Georgia. Its habitat is dense shrubbery near streams and swamps.

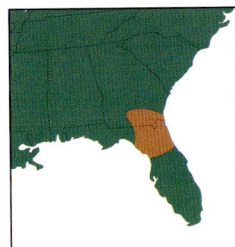

Map 41. Larger forktailed grasshopper distribution.

Melanoplus gurneyi
Strohecker

Gurney's spurthroat grasshopper

Plate 76. Gurney's spurthroat grasshopper (male).

Identification. This rare, small, shortwinged species is closely related to *Melanoplus puer* and very similar in appearance to *M. apalachicolae.* The forewings are shorter than the pronotum, elongate-oval, and widely separated dorsally. The general color is reddish brown or purplish gray above, and yellowish below. The males have behind the eye a shiny black stripe that extends across the pronotum, but this pattern is indistinct in females. This black stripe is narrow and has parallel sides on the front portion of the pronotum, widening slightly on the rear region of the pronotum. The hind femora are brownish, sometimes with black spots but not complete bands. The lower edge of the hind femora is whitish. The hind tibiae are purplish green. The males measure 13–14 mm in length, the females 17–20 mm.

In males, the furcula is short and rounded, but well developed. The cerci are symmetrically tapered throughout their length, ending in a point. The cerci are about twice as long as wide.

Similar Species. The widening of the black stripe at the rear of the side of the pronotum is important in distinguishing *M. gurneyi, M. apalachicolae, M. ordwayae* and *M. tequestae* from *M. puer* and *M. adelogyrus*. The presence of a furcula in this species serves to distinguish *M. gurneyi* from *M. tequestae* and *M. ordwayae,* which lack visible furcula. The symmetrical shape of the cerci serves to distinguish *M. gurneyi* from *M. apalachicolae,* in which the bottom edge tapers evenly throughout its length, but the top edge tapers abruptly at the base, leaving the remainder of the top edge almost straight.

Distribution and Ecology. This species is known only from the coastal area of Bay and Okaloosa counties in western Florida. It inhabits dry sand areas among oak and rosemary plants.

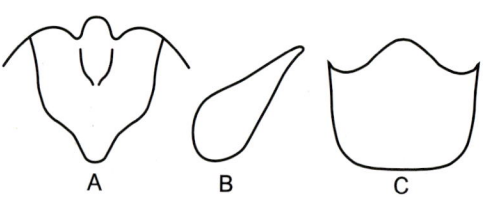

Figure 22. Supra-anal plate and furcula *(A)*, cercus *(B)*, and subgenital plate *(C)* of male *M. gurneyi.*

Map 42. Gurney's spurthroat grasshopper distribution.

Plate 77. Immodest spurthroat grasshopper (male).

Melanoplus impudicus
Scudder

Immodest spurthroat grasshopper

Identification. This is an indistinct longwinged species, hardly deserving of its common name. It is of medium size and bears long wings, grayish brown above and yellowish below. The stripe behind the eye varies from strong to weak. The forewings are marked with a modest row of small spots centrally. The outer face of the hind femora is marked with incomplete dark bands. The hind tibiae are reddish. The males measure 19–21 mm, the females 23–27 mm.

In males, the furcula consists of small structures. The cerci narrow at the middle and are only slightly expanded at the ends, tending to end in a blunt point. The outer face of the tip of the cerci is marked with a weak groove or depression.

Distribution and Ecology. The immodest grasshopper occurs through most of the eastern states, north to about New York. In Florida it is known only in the northern areas. Its habitat is dry open woodlands.

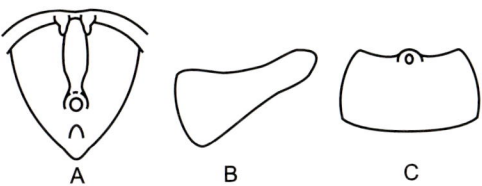

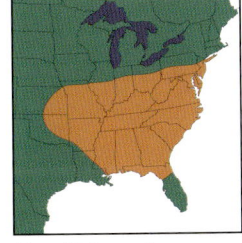

Map 43. Immodest spurthroat grasshopper distribution.

Figure 23. Supra-anal plate and furcula (A), cercus (B), and subgenital plate (C) of male *M. impudicus.*

Melanoplus indicifer Hubbell

Spinecercus spurthroat grasshopper

Identification. This small short-winged species is very similar in appearance to other *Melanoplus* species in the *puer* group. The forewings are shorter than the pronotum, oval, and widely separated above. The general color is reddish brown or gray above, and yellowish below. The males have behind the eye a shiny black stripe that extends across the pronotum, and is nearly equal in width from the front to the rear edge of the side lobe. This pattern is indistinct in females. The hind femora are dull yellow, sometimes with black spots but not complete bands. The hind tibiae are purplish green. The males measure 13–14.5 mm in length, the females about 20 mm.

In males, the furcula is not visible. The cerci are distinctive but similar to those of *M. forcipatus.* The cerci are broad at the base, with the upper and lower margins nearly parallel. Beyond the middle, however, the cerci fork into a short stubby upper projection and a long, tapering lower projection that curves downward only slightly. The supra-anal plate is less broad and less shieldlike than in many species, with the plate tapering little basally, resulting in parallel sides.

Similar Species. *Melanoplus indicifer* closely resembles *M. forcipatus*. However, the cerci of *M. forcipatus* bear a relatively prominent, broad, and strongly curved projection underneath. The cerci of *M. forcipitatus* also curve markedly inward. In contrast, the lower projection on the cerci of *M. indicifer* is less prominent, narrow, and not strongly curved. Although these two species are closely related, they do not occur in the same geographic areas of Florida, with *M. indicifer* restricted to the east coast of Florida near Palm Beach.

Distribution and Ecology. This species is found only in sandy areas along the southeastern coast of Florida, north of West Palm Beach. This portion of the state is densely populated by humans, with little habitat preserved for scrub-inhabiting animals; thus, this grasshopper may soon become extinct.

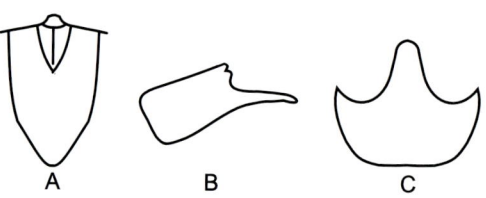

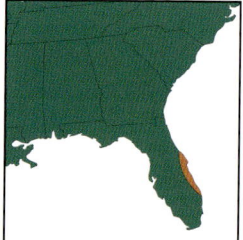

Figure 24. Supra-anal plate and furcula *(A)*, cercus *(B)*, and subgenital plate *(C)* of male *M. indicifer.*

Map 44. Spinecercus spurthroat grasshopper distribution.

Plate 78. Keeler's spurthroat grasshopper (male).

Melanoplus keeleri (Thomas)

Keeler's spurthroat grasshopper

Identification. Keeler's spurthroat is moderately large for the genus *Melanoplus* and is longwinged. It is grayish brown or reddish brown in general color, and yellowish underneath. A dark bar extends from the back of the eye onto the side lobe of the pronotum. The forewings extend to the tip of the abdomen or beyond and bear a series of small brown spots centrally. The hind femora are marked with indistinct dark bars. The hind tibiae are coral red. The males of this species measure 23–29 mm in length, the females 27–34 mm.

In males, the furcula is reduced to very small lobes. The cerci are distinctively shaped, resembling a "boot" with a rounded "toe" directed upward and a pointed "heel" directed downward.

Distribution and Ecology. Keeler's spurthroat occurs in northern Florida, and is found throughout North America east of the Rocky Mountains. Its habitat is pasture and open woods.

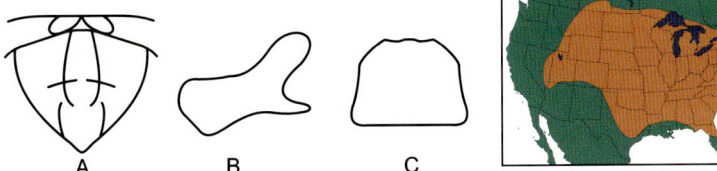

Figure 25. Supra-anal plate and furcula *(A)*, cercus *(B)*, and subgenital plate *(C)* of male *M. keeleri*.

Map 45. Keeler's spurthroat grasshopper distribution.

Melanoplus nanciae Deyrup
Ocala clawcercus grasshopper

Identification. This small shortwinged species was first described in 1996. It is fairly typical of the shortwinged, scrub-inhabiting Florida species of *Melanoplus*. It is grayish brown, but lighter below. The forewings are shorter than the prothorax, and widely separated on top of the body. A black stripe extends from the back of the eye across the prothorax, with the stripe equal in width or slightly wider at the rear region of the prothorax. The hind femora are brown, with a large dark brown spot centrally. The hind tibiae are yellowish brown at each end but bluish gray centrally. Length of males is about 14.5 mm; females measure about 19 mm.

In males, the furcula is absent. The cerci are unique and can be used to distinguish this species: they are broad at the base, dividing into a long pointed upper spine and a lower rounded lobe with a tooth. The tip of the subgenital plate is elongate and curved toward the front over the tip of the supra-anal plate.

Similar Species. Although this species superficially resembles many short-winged *Melanoplus* species, the cerci of the males distinguish them from all others. Interestingly, the female of this species also has a projection on the cercus, a feature that does not occur on other Florida shortwinged *Melanoplus*.

Distribution and Ecology. This species is known only from the Ocala National Forest in Lake County, Florida. Its habitat is poorly known, but it has been collected only from regrowth in clearcut forest following harvest of pines.

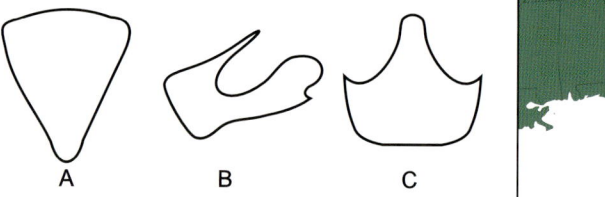 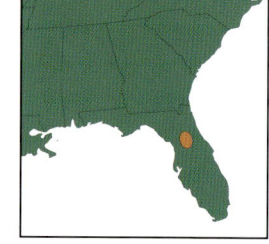

Figure 26. Supra-anal plate and furcula *(A)*, cercus *(B)*, and subgenital plate *(C)* of male *M. nanciae*.

Map 46. Ocala clawcercus grasshopper distribution.

Plate 79. Trail Ridge scrub grasshopper (male).

Melanoplus ordwayae Deyrup

Trail Ridge scrub grasshopper

Identification. This small shortwinged *Melanoplus* was not discovered until 1993. Its scientific name came from the Katherine Ordway Preserve in Putnam County, the location of its discovery. It is very similar in appearance to the numerous high pine and scrub-inhabiting *Melanoplus* species found in Florida. It is brownish on top and lighter below. The forewings are shorter than the pronotum, and widely separated on top. It bears behind the eye a dark stripe that extends across the side lobe of the pronotum. The stripe is not significantly wider at either end of the pronotum, although there may be a hint of widening at the rear. The hind femora bear a dark spot centrally. The hind tibiae are blue-gray centrally, and brownish at each end. Males measure about 15 mm in length, females about 18 mm.

In males, the furcula is not apparent. The cerci taper gradually to a point, but because the lower edge is more arched than the top, the cerci point downward.

Similar Species. Based only on its physical appearance, *M. ordwayae* is easily confused with *M. tequestae.* This latter species, however, occurs only from Orlando south to Lake Okeechobee. *Melanoplus ordwayae* occurs well north of the area supporting *M. tequestae.* Positive identification can also be made by examination of the internal genitalia. If the tip of the male abdomen in *M. tequestae* is pulled down to reveal the penis, a small forked appendage is revealed near the tip on the back side of the penis. In *M. ordwayae,* the appendage is long, originates basally, and is not forked.

Distribution and Ecology. This poorly known species has been collected only from Putnam and Clay counties, in north central Florida. It inhabits scrub areas, particularly edges of scrub oak thickets and open, white sand areas supporting nonwoody plants.

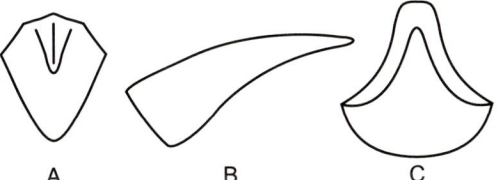

Figure 27. Supra-anal plate and furcula *(A)*, cercus *(B)*, and subgenital plate *(C)* of male *M. ordwayae.*

Map 47. Trail Ridge scrub grasshopper distribution.

Melanoplus propinquus
Scudder

Southern redlegged grasshopper

Plate 80. Southern redlegged grasshopper (male).

Identification. This long-winged species is grayish brown or yellowish brown, with a yellow abdomen. A black bar extends from the eye onto the side of the pronotum. The forewings are brownish with a row of faint spots centrally. The forewings extend to the tip of the abdomen or beyond. The hind femora are yellowish brown or greenish yellow, and lack distinct bands. The hind tibiae almost always are red, though bluish tibiae are sometimes observed. The male measures 19–26 mm in length, the female 20–29 mm.

In males, the furcula is narrow and at least one-half the length of the supra-anal plate. The cerci are wide and narrow markedly near the point of attachment and are narrow toward the point. The tip is oriented upward. The subgenital plate terminates with a U-shaped tip.

Distribution and Ecology. The southern redlegged grasshopper is abundant in weedy pastures, crop fields, along roadsides, and in other disturbed areas. Although it has been detected in most of the state, it is abundant only in northern Florida. It also is found in the coastal plain region of other southeastern states but has been replaced by *M. femurrubrum* (DeGeer) in northern areas of these states and elsewhere in North America.

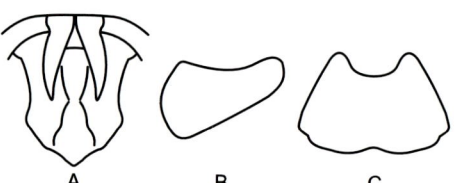

Figure 28. Supra-anal plate and furcula *(A)*, cercus *(B)*, and subgenital plate *(C)* of male *M. propinquus.*

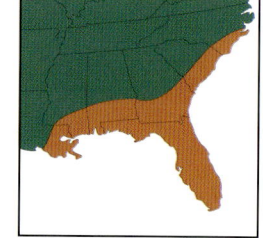

Map 48. Southern redlegged grasshopper distribution.

Melanoplus puer (Scudder)
Florida least spurthroat grasshopper

Identification. This is the smallest member of the genus *Melanoplus*, and like many of Florida's species, its wings are short. The forewings are shorter than the pronotum, elongate-oval, and widely separated on top of the body. The general color is reddish brown or purplish gray above, and yellowish below. The males display behind the eye a shiny black stripe that extends across the pronotum to the first abdominal segments, but this pattern is indistinct in females. This black stripe is very wide on the front

Plate 81. Florida least spurthroat grasshopper (male).

portion of the pronotum, narrowing markedly on the rear. The hind femora are dull yellow, sometimes with black spots but not complete bands. The hind tibiae are purplish green. The males measure 10–16 mm, the females 16–21 mm.

In males, the furcula is very short. The slender cerci taper gradually to a blunt tip, but are not completely symmetrical; the upper edge is slightly concave. The tip of the cercus is not flattened. The subgenital plate is only weakly elongated.

Similar Species. The narrowing of the black stripe on the rear of the side of the pronotum is an important character in distinguishing *Melanoplus puer* from *M. apalachicolae*, *M. gurneyi*, *M. ordwayae* and *M. tequestae*. The tip of the cerci is not blunt, as in *M. adelogyrus*.

Melanoplus puer has at least three, and perhaps five, geographic races or subspecies in Florida. By virtue of being shortwinged and inhabiting discontinuous "islands" of scrub habitat, these isolated populations have little opportunity for mating, which would result in genetic blending. Therefore, they are evolving different structural characteristics that will eventually result in recognition as separate species. The races may, in fact, be

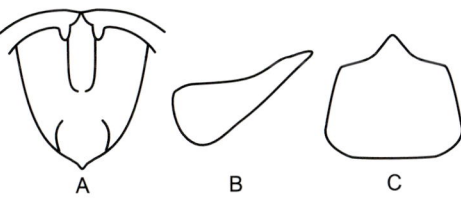

Figure 29. Supra-anal plate and furcula *(A)*, cercus *(B)*, and subgenital plate *(C)* of male *M. puer.*

sexually incompatible now because of their isolation, but have yet to evolve strongly different appearances. See the discussion on "What Is a Species?" in chapter 1 for further discussion on this topic.

Distribution and Ecology. This species is found throughout the Florida peninsula, but not elsewhere. It inhabits wire grass patches in open woods, particularly scrub and high pine habitats.

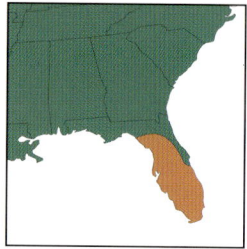

Map 49. Florida least spurthroat grasshopper distribution.

Plate 82. Pinetree spurthroat grasshopper (male).

Melanoplus punctulatus
Scudder

Pinetree spurthroat
grasshopper

Identification. This large, longwinged, grayish species is unusual in that its body and forewings bear numerous dark brown or black spots of a moderate size. The underside is reddish or yellowish. The dark bar behind the eye is indistinct. The outer face of the hind femora is marked with alternating blackish and grayish bands. The hind tibiae are reddish or gray. The males are 27–31 mm in length, the females 37–45 mm.

In males, the furcula is barely visible. The cerci are large and markedly expanded beyond the middle, appearing clublike. The subgenital plate ends with an upward extension.

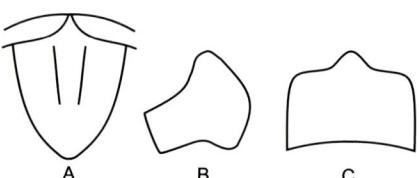

Figure 30. Supra-anal plate and furcula *(A)*, cercus *(B)*, and subgenital plate *(C)* of male *M. punctulatus*.

Distribution and Ecology. The pinetree spurthroat grasshopper is found throughout the eastern states and west to the Southwest. In Florida it is found in northern portions of the state. This poorly known species apparently inhabits cone-bearing trees such as pines and may be nocturnal. It is most often observed resting on the trunks of trees, where it blends in well with mottled bark and the moss and lichens growing on tree trunks. The female reportedly deposits her eggs within holes or crevices of dead tree trunks, a relatively uncommon habit among grasshoppers.

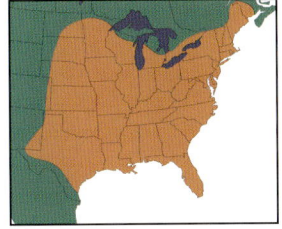

Map 50. Pinetree spurthroat grasshopper distribution.

Plate 83. Pygmy spurthroat grasshopper (male).

Melanoplus pygmaeus Davis

Pygmy spurthroat grasshopper

Identification. This small shortwinged species is poorly known. It is reddish brown above, and yellowish below. The shortened forewings are elongate-oval, and widely separated on top. As is the case with most *Melanoplus* species, a dark stripe is found behind the eye, extending onto the pronotum. The stripe is about equal in width from front to back of the side of the pronotum, but sometimes expands slightly toward the back. The hind femora are yellowish brown with three dark blotches on the upper surface. The hind tibiae are purplish blue. The male of *M. pygmaeus* measures about 14 mm in length, the female about 23 mm.

In males, the furcula is not visible. The cerci are pinched near the middle and elbowed, with the tip turned upward. The tip is broadly rounded and flattened, and slightly concave or grooved.

Similar Species. The presence of a recessed or grooved area on the tip of the cerci cause *M. pygmaeus* to resemble *M. rotundipennis.* The absence of a furcula and a pallium serve to distinguish this species from *M. rotundi-pennis.*

Distribution and Ecology. This species has been collected only in western Florida. It is found in high pine and scrub habitats.

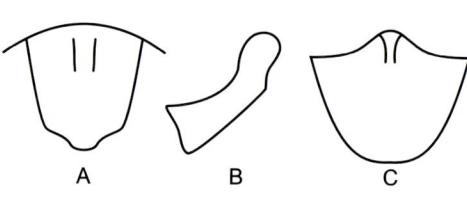

Figure 31. Supra-anal plate and furcula *(A)*, cercus *(B)*, and subgenital plate *(C)* of male *M. pygmaeus.*

Map 51. Pygmy spurthroat grasshopper distribution.

Plate 84. Oak spurthroat grasshopper (male).

Melanoplus querneus
Rehn and Hebard

Oak spurthroat grasshopper

Identification. This heavybodied *Melanoplus* with forewings of intermediate length is flightless. The forewings normally extend two-thirds to three-fourths the length of the abdomen. The body and forewings are brown with yellowish or grayish markings above, yellowish green below. A dark stripe extends from the eye onto the pronotum, but sometimes is relatively indistinct. The outer face of the hind femora is marked with two dark bands. The hind tibiae are reddish. The males measure 22–27 mm in length, the females 28–40 mm.

In males, the furcula is greatly reduced or not apparent. The cerci are large and expanded beyond the middle, especially on the upper side.

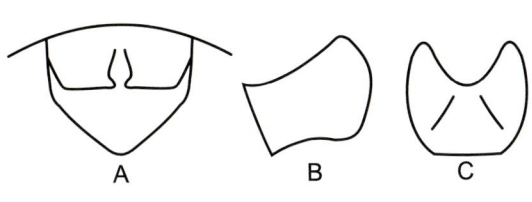
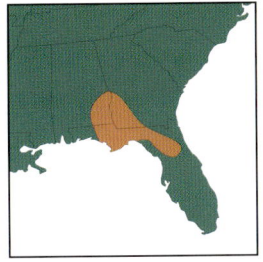

Figure 32. Supra-anal plate and furcula *(A)*, cercus *(B)*, and subgenital plate *(C)* of male *M. querneus*.

Map 52. Oak spurthroat grasshopper distribution.

Distribution and Ecology. Found in northern Florida and adjacent states, this grasshopper inhabits the undergrowth and margins of woodlands. It seems to prefer moist areas or taller vegetation.

Melanoplus rotundipennis (Scudder)

Roundwinged spurthroat grasshopper

Plate 85. Roundwinged spurthroat grasshopper (male).

Identification. This is the most common of the shortwinged *Melanoplus* species in Florida. It is a small grasshopper, but about average in size for the shortwinged species. It is reddish brown on top and yellowish below. A dark stripe is found behind the eye, and it extends over the pronotum and onto the abdomen in males, but only to about the middle of the pronotum in females. The hind femora are yellowish or brownish, often with two bands crossing them. The hind tibiae are bluish. The forewings are not really round, despite the common name, but they are only slightly elongate-oval. The forewings are widely separated on top of the body. Males of this grasshopper measure about 13.5–17.5 mm in length, females 17–23 mm.

In males, the furcula is very short, consisting of rounded lobes. The cerci are pinched near the middle, and slightly widened and flattened or shallowly grooved at the tip. The most striking feature in the male of this species is the enlarged pallium, an erect conical structure at the tip of the supra-anal plate. This structure protrudes from the tip of the upper surface of the abdomen, and is an important character for distinguishing this species from the other shortwinged *Melanoplus*.

Similar Species. The presence of a pallium, an erect conical structure at the tip of the supra-anal plate, distinguishes this species from most other short-winged *Melanoplus* species. Only *M. withlacoocheensis* also possesses the enlarged pallium, but *M. withlacoocheensis* is easily distinguished because the tips of the cerci have a small tooth below and are swollen, appearing bulbous or knob-like when viewed from above.

Distribution and Ecology. This species occurs only in northern Florida and southern Georgia. Within Florida it occurs south to about Orlando and Lakeland. It is common in a number of dry and moderately moist habitats, including dry hammocks, flatwoods, high pine, and scrub oak. It is particularly common along edges of woods, and sometimes ventures out into old fields.

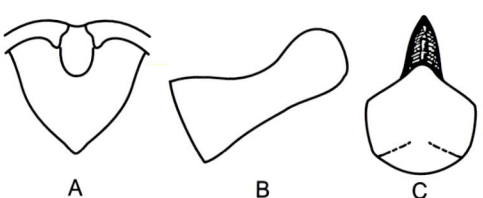

Figure 33. Supra-anal plate and furcula *(A)*, cercus *(B)*, and subgenital plate *(C)* of male *M. rotundipennis.*

Map 53. Roundwinged spurthroat grasshopper distribution.

Melanoplus sanguinipes (Fabricius)
Migratory grasshopper

Identification. Migratory grasshopper is grayish brown. A black stripe usually extends from the eye onto the side of the pronotum. The forewings are long, brownish, and bear a row of dark brown spots centrally. The forewings extend to the tip of the abdomen or beyond. The hind femora usually have two slanted broad dark bands. The hind tibiae normally are red, but sometimes blue. The male measures 19–24 mm, the female 18–29 mm.

Plate 86. Migratory grasshopper (male).

In males, the furcula is slender, V-shaped, and measures about one-fourth to one-third the length of the supra-anal plate. The cerci are compact, about twice as long as broad, and rounded at the tip. The tip of the subgenital plate is extended; when viewed from above it is clearly notched in the middle.

Distribution and Ecology. The migratory grasshopper occurs throughout North America. In western states it sometimes attains very high and damaging densities. At high densities a behavioral change occurs wherein the grasshoppers become gregarious, moving as a group. During such times the grasshoppers may disperse long distances, resulting in the common name "migratory grasshopper." In Florida, the migratory grasshopper does not become excessively abundant, and causes little damage. Its distribution is restricted to northern Florida. The favored habitat of the migratory grasshopper is weedy pastures, crops, and similar disturbed areas where annual weeds are abundant.

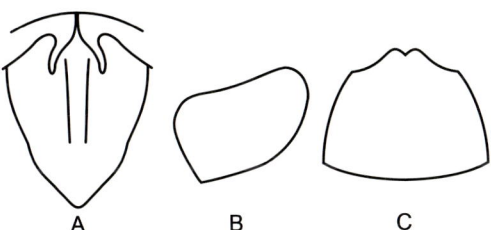

Figure 34. Supra-anal plate and furcula (A), cercus (B), and subgenital plate (C) of male M. sanguinipes.

Map 54. Migratory grasshopper distribution.

Plate 87. Lesser forktailed grasshopper (male).

Melanoplus scapularis
Rehn and Hebard

Lesser forktailed grasshopper

Identification. This short-winged species is not unlike the very common *Melanoplus rotundipennis* in general appearance. It is reddish brown above and yellowish below. The black bar behind the eye extends across the pronotum onto the abdomen in both sexes. The bar is equal in width, or expanding slightly at the back edge of the pronotum. The short oval forewings overlap, or are only narrowly separated on top. The outer face of the hind femora is generally brown, without bars crossing it. The hind tibiae are bluish gray or brownish. The males measure 15–17.5 mm in length, the females 19–22 mm.

In males, the furcula is not visible. The most distinguishing characteristic of this species is the shape of the cerci. The cerci are enlarged, expanding markedly from the base and usually into a broadly rounded projection on top and a more pointed projection below. Sometimes the upper projection is also pointed.

Similar Species. The overlapping forewings of this species cause it to resemble *M. davisi, M. scudderi,* and *M. strumosus.* Although similar to these other small *Melanoplus* species, *M. scapularis* is easily distinguished based on the shape of the forked cerci.

Distribution and Ecology. This species is known from northern Florida and adjacent states. Its habitat is scrub oak woods or among low-growing bushes under pines on sandy soil.

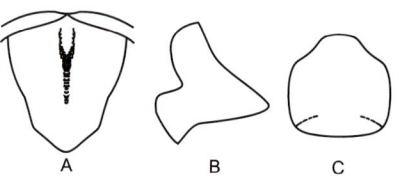

Figure 35. Supra-anal plate and furcula *(A)*, cercus *(B)*, and subgenital plate *(C)* of male *M. scapularis.*

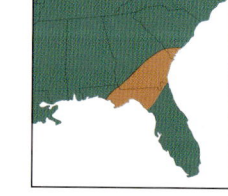

Map 55. Lesser forktailed grasshopper distribution.

Melanoplus scudderi (Uhler)

Scudder's spurthroat grasshopper

Identification. This shortwinged species is of medium size, and brownish or reddish brown in general body color. The oval or elongate-oval forewings overlap, or are only slightly separated, on top of the body. In length, the forewings vary from less than the length of the pronotum to slightly longer. The dark stripe normally found behind the eye in *Melanoplus* species may be present or weak in both sexes. The hind femora lack bands crossing their outer faces, but two dark spots may be present on top. The hind tibiae are red. The males measure 14–18.5 mm in length, females 22–24 mm.

In males, the furcula consists of very small triangular structures, and sometimes is not apparent. The cerci taper from the base to a broadly rounded point, usually curving upward. The cerci are slightly concave or grooved toward the tip.

Similar Species. *Melanoplus scudderi* is similar to *M. davisi,* with both species having cerci that turn upward into blunt tips. However, the furcula of *M. davisi* appears to be very small rounded appendages, whereas in *M. scudderi* they are pointed. Also, the tips of the cerci are much wider in *M. scudderi* than in *M. davisi.* Other shortwinged species that have overlapping forewings include *M. scapularis* and *M. strumosus,* but these are easily distinguished by the markedly different shape of their cerci.

Distribution and Ecology. Scudder's spurthroat grasshopper is widely distributed in the eastern United States, west to Nebraska and Texas. In Florida its distribution is limited to northern areas of the state. Its habitat is among low-growing oaks and grasses in oak and longleaf pine woods growing on sandy soil.

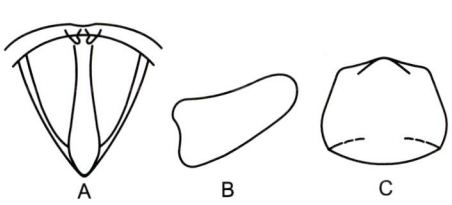

Figure 36. Supra-anal plate and furcula *(A)*, cercus *(B)*, and subgenital plate *(C)* of male *M. scudderi.*

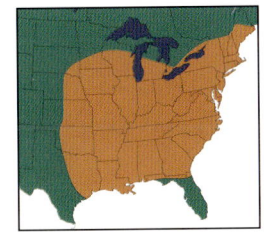

Map 56. Scudder's spurthroat grasshopper distribution.

Plate 88. Swollen spurthroat grasshopper (male).

Melanoplus strumosus
Morse

Swollen spurthroat grasshopper

Identification. A small, shortwinged species, this grasshopper superficially resembles many other species in the *Melanoplus puer* group, but has some easily observed distinguishing characteristics. Its general color is reddish brown above and whitish below. The dark stripe behind the eye crosses the pronotum and extends onto the abdomen in males, but is indistinct in females. The oval forewings meet, or are only slightly separated, on top of the body. The yellowish hind femora may have two brownish bands crossing them, but often the bands are weak or lacking. The hind tibiae are bluish. The males measure 15–17 mm in length, the females 17.5–26 mm.

In males, the furcula tapers markedly at the base, and extends over about one-third to one-half the length of the supra-anal plate. The shape of the furcula apparently is the basis for the common name. The cerci are long and slender, pinched at the middle, and bear a point at the tip that is oriented downward.

Similar Species. The tendency of the forewings to meet on top of the body is a character shared with *Melanoplus davisi, M. scapularis,* and *M. scudderi,* but the slender shape of the cerci distinguishes *M. strumosus* from these similar grasshoppers.

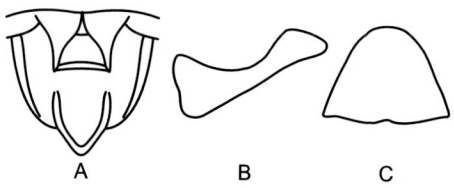

Figure 37. Supra-anal plate and furcula *(A)*, cercus *(B)*, and subgenital plate *(C)* of male *M. strumosus.*

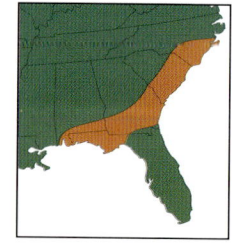

Map 57. Swollen spurthroat grasshopper distribution.

Distribution and Ecology. *Melanoplus strumosus* is known throughout the southeastern states. In Florida, however, it is known only in the northern areas. It inhabits pine woods with an oak understory, particularly among oak sprouts.

Melanoplus symmetricus Morse

Symmetrical spurthroat grasshopper

Plate 89. Symmetrical spurthroat grasshopper (male).

Identification. This large, heavybodied, longwinged species is brownish yellow. There is sometimes a dark bar behind the eye that extends onto the pronotum. The hind femora generally are yellowish, but sometimes brownish. The hind tibiae are dull red. The males measure 28–30 mm in length, the females 31–36 mm.

In males the furcula is not visible. The cerci are large and flat. They are pinched at the middle and expanded upward and downward at the tip. The upper lobe may be more expanded, or equivalent to the lower lobe.

Similar Species. *Melanoplus symmetricus* is closely related to *M. furcatus,* and may eventually prove to be a form of that species. Presently, they are distinguished by the shape of the cercus; in *M. symmetricus,* the cercus is symmetrical, expanding uniformly upward and downward. In *M. furcatus,* the expansion is V-shaped, and not symmetrical.

Distribution and Ecology. This species is known only from western Florida. Its habitat is dense shrubbery near streams and swamps.

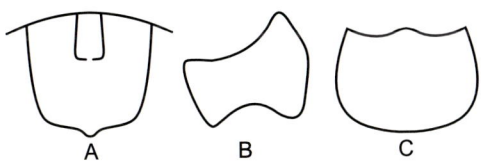

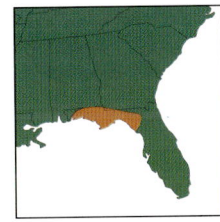

Figure 38. Supra-anal plate and furcula *(A)*, cercus *(B)*, and subgenital plate *(C)* of male *M. symmetricus.*

Map 58. Symmetrical spurthroat grasshopper distribution.

Melanoplus tepidus Morse
Southern obovatewinged grasshopper

Identification. This medium-sized, shortwinged species is brownish on top and yellowish below. The wings are oval or egg-shaped, providing the basis for the common name. A shiny black stripe extends from the eye back across the pronotum; it is about equal in width throughout its length. Below this dark stripe the face and pronotum are ivory white. The forewings are elongate, shorter than the pronotum, and separated above. The hind femora bear two dark bands. The hind tibiae are grayish. The males measure about 16–20 mm in length, the females 23–29 mm.

In males, the furcula is wide at the base, tapering rapidly to a point. The furcula is about one-fourth the length of the supra-anal plate. The cerci are broad at their bases, constricted near the middle, and expanded into bluntly rounded tips that are oriented upward. The tip is concave or grooved. The subgenital plate bears weak evidence of a blunt extension below the tip of the abdomen.

Distribution and Ecology. This poorly known species has been recorded in Florida only from Liberty County in the western Panhandle, but may occur elsewhere. It also has been collected in Alabama. Its habitat is undescribed. *Melanoplus tepidus* is similar to a wide-ranging, more northern species, *M. obovatipennis* (Blatchley); *M. tepidus* may eventually be found to be a form of this species.

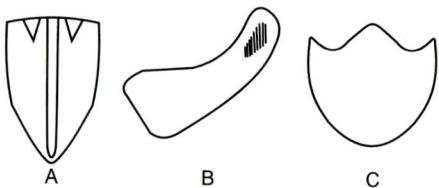

Figure 39. Supra-anal plate and furcula (A), cercus (B), and subgenital plate (C) of male *M. tepidus.*

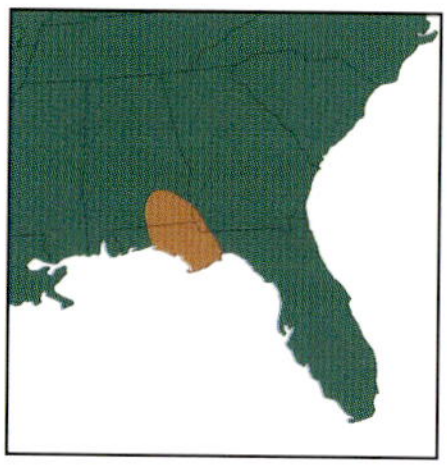

Map 59. Southern obovatewinged grasshopper distribution.

Melanoplus tequestae Hubbell
Tequesta spurthroat grasshopper

Identification. A small short-winged species, *Melanoplus tequestae* is closely related to *M. puer,* and very similar in appearance to *M. apalachicolae* and *M. ordwayae.* The forewings are shorter than the pronotum,

Plate 90. Tequesta spurthroat grasshopper (male).

elongate-oval, and widely separated on top of the insect. The general color is reddish brown or purplish gray above, and yellowish below. This species is more compact and heavybodied than *M. apalachicolae,* however. The males have a shiny black stripe behind the eye that extends across the pronotum, but this pattern is indistinct in females. This black stripe is narrow and has parallel sides on the front portion of the pronotum, widening slightly at the rear of the pronotum. The hind femora are dull yellow, sometimes with black spots but not complete bands. The hind tibiae are purplish green. The males measure 11.5–15 mm in length, the females 17–20.5 mm.

In males the furcula is not visible. The cerci are tapered gradually from base to tip, but usually curved more underneath than on top, causing the tip of the cerci to be pointed downward. In the southern range of this species some males display cerci that are symmetrical, not curving downward. The cerci are about three times as long as wide.

Similar Species. The width of the black stripe on the side lobe of the pronotum is an important character in distinguishing *Melanoplus tequestae, M. ordwayae, M. gurneyi,* and *M. apalachicolae* from *M. puer* and *M. adelogyrus.* The absence of a furcula serves to distinguish *M. tequestae* from *M. apalachicolae* and *M. gurneyi,* which have short but apparent furcula. This species is easily confused with *M. ordwayae,* with which it shares a

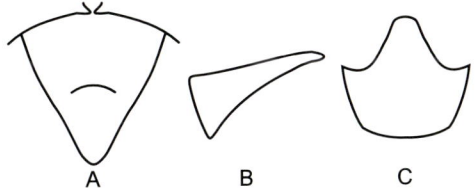

Figure 40. Supra-anal plate and furcula *(A),* cercus *(B),* and subgenital plate *(C)* of male *M. tequestae.*

tendency for the male cerci to be curved downward, but *M. ordwayae* is found only north of Orlando. Positive identification can also be made by examining the internal genitalia. If the tip of the male abdomen in *M. tequestae* is pulled down to reveal the penis, a small forked appendage is revealed near the end of the penis. In *M. ordwayae* the appendage is long, originates at the base of the penis, and is not forked.

Distribution and Ecology. This species is named for a tribe of Native Americans that inhabited the lower east coast of Florida at the time of Spanish exploration. This species is known only from the central sand ridge area of central Florida, bounded by the Orlando area in the north and the Lake Okeechobee area in the south. As is the case with most other short-winged *Melanoplus* species, the principal habitat is open scrub oak.

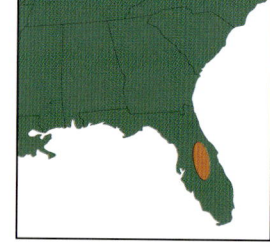

Map 60. Tequesta spurthroat grasshopper distribution.

Plate 91. Withlacoochee grasshopper (male).

Melanoplus withlacoocheensis Squitier and Deyrup

Withlacoochee grasshopper

Identification. This small shortwinged species was discovered inhabiting the southern portion of the Brooksville Ridge, in west central Florida in 1998. The grasshopper derives its name, *withlacoocheensis,* from its discovery in and near the Withlacoochee State Forest. It is grayish brown on top and cream colored underneath. The oval forewings are widely separated on top. A black stripe extends from behind the eye across the side of the pronotum. The hind legs are grayish brown on top, but fade into a cream color below. The hind tibiae are purple to bluish. Males of this grasshopper measure 15.4–16.5 mm in length, the females 20–29.5 mm.

In males, the furcula is small and reduced to two rounded lobes. The cerci are swollen or bulbous at the ends, with a small tooth underneath. This species has a large and erect pallium that houses the genitalia.

Similar Species. This species easily could be mistaken for *M. rotundipennis,* from which it is derived. However, the presence of the swollen cerci and the tooth under the tip of the cerci distinguish *M. withlacocheensis* from *M. rotundipennis.* Internally, the larger size and snake-like shape of the penis in *M. withlacoocheensis* further aid in differentiation.

Distribution and Ecology. This species has been collected only on the southern portion of the Brooksville Ridge in Citrus and Hernando counties. It is numerous in high pine habitats containing open areas. So far *M. rotundipennis* has been found only along the eastern edge of the range of *M. withlacoocheensis.*

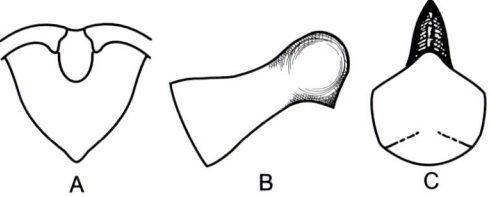 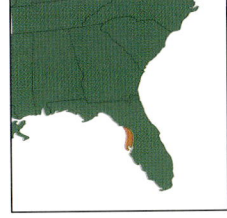

Figure 41. Supra-anal plate and furcula *(A),* cercus *(B),* and subgenital plate *(C)* of male *M. withlacoocheensis.*

Map 61. Withlacoochee grasshopper distribution.

Paroxya atlantica
Scudder

Atlantic grasshopper

Plate 92. Atlantic grasshopper (male).

Identification. The Atlantic grasshopper is usually yellowish brown, but sometimes tends toward maroon or red. Its face is moderately slanted. Found behind the eye is a black stripe that fades near the middle of the pronotum, as is common among *Melanoplus* species. Indeed, it easily is confused with *Melanoplus* spp., but its antennae and pronotum are exceedingly long. In males, the antennae are about twice the length of the pronotum. Also in males, the pronotum is elongate, about twice as long as wide. In *Melanoplus* the antennae and pronotum are shorter. The hind tibiae are bright blue or greenish blue. In males, the cerci are long, slender, constricted at the middle, strongly incurved, with the tip flattened and broadly rounded. The furcula is short or barely visible. Males measure 16–24 mm in length, females 22–28 mm.

Similar Species. The length of the antennae and pronotum serve to distinguish this species from *Melanoplus* species. The moderate size of *P. atlantica* serves to distinguish it from the similar, but larger, *P. clavuliger*. Also, the antennae of male *P. atlantica* are shorter than the hind femora, whereas in male *P. clavuliger* the antennae are longer than the hind femora.

Distribution and Ecology. The Atlantic grasshopper is found throughout Florida and most of the eastern United States. It inhabits wet areas, and is particularly common on the vegetation around ponds, swamps, and in coastal salt marshes.

Map 62. Atlantic grasshopper distribution.

Plate 93. Olivegreen swamp grasshopper (male).

Paroxya clavuliger
(Serville)

Olivegreen swamp grasshopper

Identification. This species is greenish to greenish black, closely resembling *Paroxya atlantica* in most respects. As in *P. atlantica*, *P. clavuliger* is marked with a dark stripe extending from the eye onto the pronotum, but unlike in *P. atlantica*, the stripe usually continues to the hind edge of the pronotum. *Paroxya clavuliger* is easily confused with *Melanoplus* species, but its antennae and pronotum are very long. In males, the antennae are about twice the length of the pronotum and longer than the hind femora. Also in males, the pronotum is elongate, about twice as long as wide. In *Melanoplus* the antennae and pronotum are shorter. This species has blue or bluish green hind tibiae. In males, the cerci are long, slender, constricted at the middle, strongly incurved, and with the tip flattened and broadly rounded. The furcula is evident, measuring one-fourth to one-third the length of the supra-anal plate. The males measure 20–27 mm in length, the females 29–40 mm.

Similar Species. The lengths of the antennae and pronotum serve to distinguish this species from *Melanoplus* species. The moderate size of *P. atlantica* serves to distinguish it from the similar, but larger, *P. clavuliger*. Also, the antennae of male *P. atlantica* are shorter than the hind femora, whereas in male *P. clavuliger* the antennae are longer than the hind femora.

Distribution and Ecology. *Paroxya clavuliger* occurs throughout Florida, and is widespread in the eastern United States. It inhabits wet areas, and is normally associated with the edges of ponds, freshwater marshes, and coastal salt marshes.

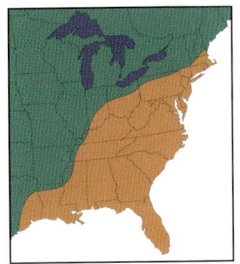

Map 63. Olivegreen swamp grasshopper distribution.

Schistocerca alutacea (Harris)

Rusty grasshopper

Plate 94. Rusty grasshopper (male).

Identification. The rusty grasshopper is highly variable in appearance. Its general body color often is golden brown or rusty brown which, of course, is the basis of the common name. It may also be olive with a yellowish stripe running the length of the body from the head to the tip of the

Plate 95. Rusty grasshopper (female).

forewings, as is found in *Schistocerca obscura,* but *S. alutacea* generally is smaller. Olive-colored individuals of *S. alutacea* are easily confused with *S. obscura,* and the best approach to distinguish between the two species is to examine the tip of the male abdomen. In *S. obscura* the notch of the male's subgenital plate is V-shaped, whereas in *S. alutacea* it is U-shaped. Occasionally, females lack the stripe and instead bear indistinct brownish spots on the forewings. The hind tibiae of the rusty grasshopper are brownish, with yellow spines bearing dark tips. In body length the males of *S. alutacea* measure 30–40 mm, the females 43–54 mm.

Similar Species. The rusty grasshopper may sometimes be confused with *Schistocerca obscura,* but *S. obscura* tends to be larger, and the notch of the male's subgenital plate is V-shaped. Also, the base color of the hind tibiae of *S. obscura* is purplish or blackish rather than the brown of *S. alutacea.* The rusty grasshopper should not be confused with the brownish *S. damnifica,* because it is larger than this latter species.

Distribution and Ecology. The rusty grasshopper is found throughout Florida, and all except the northernmost regions of the United States. It is found commonly in open woods, especially sandy areas where scrub oak is abundant. It may also occur in pastures and the margins of wooded areas.

Figure 42. Subgenital plates showing notches.

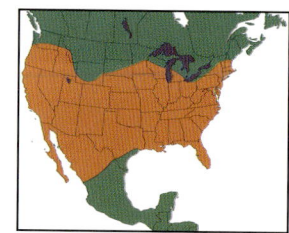

Map 64. Rusty grasshopper distribution.

Schistocerca americana (Drury)

American grasshopper

Plate 96. American grasshopper (male).

Plate 97. American grasshopper (nymphs).

Identification. This is a large, strong-flying species that is numerous enough, on occasion, to become a crop and tree pest. The American grasshopper is normally brownish or yellowish brown, with lighter and darker areas. It usually bears a creamy white stripe on top, extending from the front of the head to the tip of the forewings. Immediately after molting to the adult stage, this grasshopper is pinkish or reddish, but after a week or so the typical brown or yellow-brown color is acquired. The forewings, which extend well beyond the tip of the abdomen, bear large, dark brown spots. The hind tibiae are red. The male measures 39–52 mm in length, the female 48–68 mm.

Distribution and Ecology. The American grasshopper occurs throughout Florida and the eastern United States. It is unusual in having two generations annually, one usually in April through June and another beginning in August or September. The adults overwinter, and are active in the winter whenever it is warm and sunny. When weather and food conditions allow the American grasshopper to become abundant, behavioral changes become noticeable. Specifically, nymphs and adults become gregarious, moving in unison and dispersing in swarms. Under such conditions they can be very damaging to crops. The habitat of the American grasshopper is open fields, and open oak and pine woodlands. The American grasshopper feeds on a wide variety of grasses, forbs,

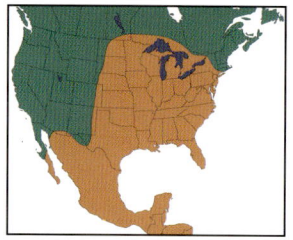

shrubs, and trees. When disturbed, it often flies into trees or a considerable distance from the source of disturbance.

Map 65. American grasshopper distribution.

Plate 98. Rosemary grasshopper (male).

Schistocerca ceratiola
Hubbell and Walker

Rosemary grasshopper

Identification. This species is mottled gray and brown with green on the abdomen. A faint pale stripe occurs on top of the head and pronotum, extending weakly along the forewings. The underside is markedly paler. The hind tibiae are red or brown. This species is quite slim in general appearance. Males of rosemary grasshopper measure 28–32 mm in length, females 36–40 mm.

Distribution and Ecology. Despite its moderately large size, the rosemary grasshopper escaped notice until it was discovered in central Florida in 1928. It escaped earlier detection owing to (1) its restricted range; it occurs only in Florida's central and southeastern sandy ridges, (2) its restricted diet; it feeds only on Florida rosemary, *Ceratiola ericoides,* and (3) its restricted period of activity; it is active only at night. It hides deep within Florida rosemary bushes (not related to the rosemary used in cooking), where it is effectively camouflaged during the daylight hours, and moves to the surface of the bushes at night. This species occurs only in Florida. Although not uncommon in certain natural areas, there is concern about its survival because its continued existence depends on availability of rapidly disappearing habitats: scrub and high pine.

Map 66. Rosemary grasshopper distribution.

Schistocerca damnifica (Saussure)

Mischievous grasshopper

Plate 99. Mischievous grasshopper (male).

Identification. The mischievous grasshopper is reddish brown, usually with a narrow brown line along the head and pronotum. It lacks the pronounced yellowish line commonly appearing along the back of the other *Schistocerca* species. The forewings extend beyond the tip of the abdomen, but to a lesser degree than the other *Schistocerca* species in Florida. *Schistocerca damnifica* is a relatively small member of the genus, males measuring 25–29 mm in length, females 37–46 mm.

Similar Species. The lack of a light stripe along the back, and its small size, serve to distinguish *D. damnifica* from other *Schistocerca* species.

Distribution and Ecology. The habitat of the mischievous grasshopper is old fields and open woodlands; in the latter environment it can be quite common at times. It occurs throughout Florida and the eastern United States except for New England and the Great Lakes region.

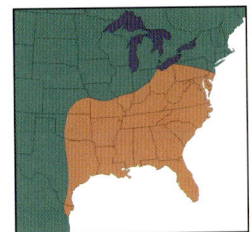

Map 67. Mischievous grasshopper distribution.

Schistocerca obscura (Fabricius)

Obscure birdwing grasshopper

Plate 100. Obscure birdwing grasshopper (female).

Identification. The obscure birdwing grasshopper is green, with olive green forewings and usually a pale yellow stripe, extending from the front of the head to the tip of the forewings. Occasionally females lack the stripe and instead bear indistinct brownish spots on the forewings. The hind tibiae are blackish purple with yellow, black-tipped spines. The obscure birdwing grasshopper is a large species; males measure 36–45 mm in length, females 50–65 mm.

Similar Species. The obscure birdwing grasshopper can be confused with *Schistocerca alutacea,* although it usually is considerably larger than this latter species. To distinguish between the two species, examine the tip of the male abdomen. In *S. obscura* the notch of the male's subgenital plate is V-shaped, whereas in *S. alutacea* it is U-shaped. The large size and V-shaped notch of the male's subgenital plate serve to distinguish *S. obscura* from *S. alutacea.*

Distribution and Ecology. The obscure birdwing grasshopper is found throughout Florida, and occurs widely in the eastern United States. Its preferred habitat is fields and open woodlands.

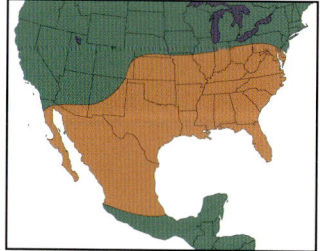

Map 68. Obscure birdwing grasshopper distribution.

Stenacris vitreipennis (Marschall)

Glassywinged toothpick grasshopper

Plate 101. Glassywinged toothpick grasshopper.

Identification. This very slender, elongate grasshopper has a distinctly pointed head and flattened, sword-shaped antennae. Thus, this grasshopper superficially resembles species from the subfamily Gomphocerinae, but can be distinguished by the presence of the spur or spine between the front legs. The glassywinged toothpick grasshopper is green to brownish green, and usually has a dark or pale line on each side extending from the eye to the base of the front legs. The length of the head is shorter than the pronotum. The length of male grasshoppers is 24–26 mm, whereas in females it is 27–29 mm.

Similar Species. *Stenacris vitreipennis* is easily confused with *Leptysma marginicollis,* but in *L. marginicollis* the head is as long as, or longer than, the pronotum whereas in *S. vitreipennis* the head is shorter than the pronotum. The antennal segments, although flattened, are not nearly as wide as in *L. marginicollis.*

Distribution and Ecology. *Stenacris vitreipennis* is known throughout Florida except for the Keys, and in most other southeastern states. The habitat of the glassywinged toothpick grasshopper is semiaquatic vegetation such as cattails and pickerelweed. This species flies readily if disturbed, tending to alight on emergent vegetation, where it dodges to the side of the plant opposite the source of disturbance. Thus, in all respects, the glassywinged toothpick grasshopper is similar to the cattail toothpick grasshopper, *Leptysma marginicollis.*

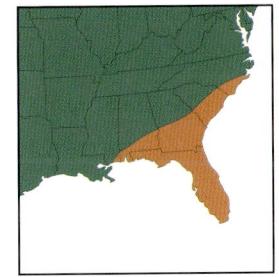

Map 69. Glassywinged toothpick grasshopper distribution.

Lubber Grasshoppers

Subfamily ROMALEINAE

This group is sometimes considered to be a separate family, Romaleidae. It differs from the other subfamilies in having an immovable spine at the tip of the hind tibiae. Lubber grasshoppers bear a spine between the front legs, as is found in spurthroated grasshoppers, subfamily Cyrtacanthacridinae. Lubber grasshoppers are large, colorful, and usually bear short wings. The shape of the head, though variable, is usually broadly rounded. The hind femora are enlarged. When disturbed, lubber grasshoppers may hiss and spread their wings. Both the front wings and hind wings are brightly colored.

Only a few species occur in North America, although many are native to South America. Just one species is known to live in the eastern United States:

Romalea
R. microptera (**Beauvois**)

Plate 102. Eastern lubber grasshopper (female).

Romalea microptera (Beauvois)

Eastern lubber grasshopper

Identification. This species is also sometimes known as *Romalea guttata* (Houttuyn). Despite the confusion in the scientific literature concerning the correct name, Floridians have little trouble recognizing this insect. It is undoubtedly the best-known species of grasshopper in Florida, and one of the most readily recognized insects.

The nymphs are mostly black with a narrow yellow stripe along the back, and red on the head and front legs. Their color pattern is distinctly different from the adult stage, so they commonly are mistaken for a different species. Young tend to be gregarious and dispersive, moving in small groups. This commonly brings them into contact with people and gardens, accounting for their familiarity. On occasion they are abundant enough to damage citrus or vegetables. They commonly seek out and defoliate amaryllis and related plants in flower gardens.

Adults are colorful, but the color pattern varies within the state. In northern Florida, this grasshopper is mostly black but well marked with yellow. The lubbers in southern Florida, however, are mostly yellow but bear red and black markings and red on the forewings. Intermediate forms also exist. Adults have small wings measuring no more than two-thirds the length of the abdomen, and are flightless. They attain a large size, males measuring 43–55 mm in length and females measuring 50–70 mm.

Distribution and Ecology. The eastern lubber grasshopper is found throughout the state, and is common in other southeastern states. There is but a single generation annually, but either nymphs or adults are present throughout most of the year in the southern portions of Florida and all except the coldest months in northern Florida. Both sexes stridulate by rubbing the forewing against the hind wing. When alarmed, lubbers spread their wings, hiss, and secrete foul-smelling froth from their spiracles, tiny holes on the sides of their abdomen through which they "breathe." Lubbers prefer to inhabit low, moist areas of dense undergrowth including wet hammocks with moderately dense overstory, but as they mature they disperse widely and can be found in nearly all habitats.

The eastern lubber grasshopper has wide-ranging tastes. It eats the leaves of at least 26 species from 15 plant families containing shrubs, herbs, broadleaf weeds, and grasses. It is reported to display preference for pokeweed, *Phytolaca americana*; tread-softly, *Cnidoscolus stimmulosus;* pickerel weed, *Pontederia cordata;* lizard's tail, *Saururus* sp.; sedge, *Cyperus;* and arrowhead, *Sagittaria* spp. Though the preferred habitat seems to be low, wet areas in pastures and woods and along ditches, lubbers disperse long distances during the nymphal period. They are gregarious but flightless, their migrations sometimes bringing large numbers into contact with crops where they damage vegetables, fruit trees, and ornamental plants. Thus, despite an apparent preference for moist areas, they can be found in nearly all habitats.

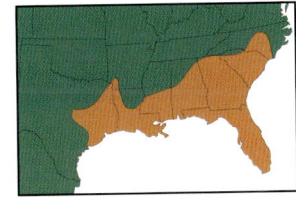

Map 70. Eastern lubber grasshopper distribution.

Appendix

Checklist of Florida Grasshoppers

Achurum
    A. carinatum (F. Walker) · longheaded toothpick grasshopper
Amblytropidia
    A. mysteca (Saussure) · brown winter grasshopper
Aptenopedes
    A. apterae Scudder · wingless Florida grasshopper
    A. sphenarioides Scudder · linearwinged grasshopper
Arphia
    A. granulata (Saussure) · southern yellowwinged grasshopper
    A. sulphurea (Fabricius) · sulfurwinged grasshopper
    A. xanthoptera (Burmeister) · autumn yellowwinged grasshopper
Chortophaga
    C. australior (Rehn and Hebard) · southern greenstriped grasshopper
Dichromorpha
    D. elegans (Morse) · elegant grasshopper
    D. viridis (Scudder) · shortwinged green grasshopper
Dissosteira
    D. carolina (Linnaeus) · Carolina grasshopper
Eotettix
    E. palustris Morse · swamp eastern grasshopper
    E. pusillus Morse · little eastern grasshopper
    E. signatus Scudder · handsome Florida grasshopper
Eritettix
    E. obscurus (Scudder) · obscure slantfaced grasshopper
Gymnoscirtetes
    G. morsei Hebard · Morse's wingless grasshopper
    G. pusillus Scudder · little wingless grasshopper

Hesperotettix

    H. floridensis Morse · Florida purplestriped grasshopper

    H. osceola Hebard · Osceola's grasshopper

    H. viridis (Thomas) · meadow purplestriped grasshopper

Hippiscus

    H. ocelote (Saussure) · wrinkled grasshopper

Leptysma

    L. marginicollis (Serville) · cattail toothpick grasshopper

Melanoplus

    M. adelogyrus Hubbell · St. Johns' spurthroat grasshopper

    M. apalachicolae Hubbell · Apalachicola spurthroat grasshopper

    M. bispinosus Scudder · twospined spurthroat grasshopper

    M. davisi (Hebard) · Davis' oak grasshopper

    M. forcipatus Hubbell · toothcercus spurthroat grasshopper

    M. furcatus Scudder · larger forktailed grasshopper

    M. gurneyi Strohecker · Gurney's spurthroat grasshopper

    M. impudicus Scudder · immodest spurthroat grasshopper

    M. indicifer Hubbell · spinecercus spurthroat grasshopper

    M. keeleri (Thomas) · Keeler's spurthroat grasshopper

    M. nanciae Deyrup · Ocala clawcercus grasshopper

    M. ordwayae Deyrup · Trail Ridge scrub grasshopper

    M. propinquus Scudder · southern redlegged grasshopper

    M. puer (Scudder) · Florida least spurthroat grasshopper

    M. punctulatus Scudder · pinetree spurthroat grasshopper

    M. pygmaeus Davis · pygmy spurthroat grasshopper

    M. querneus Rehn and Hebard · oak spurthroat grasshopper

    M. rotundipennis Scudder · roundwinged spurthroat grasshopper

    M. sanguinipes (Fabricius) · migratory grasshopper

    M. scapularis Rehn and Hebard · lesser forktailed grasshopper

    M. scudderi (Uhler) · Scudder's spurthroat grasshopper

    M. withlacoocheensis Squitier and Deyrup · Withlacoochee grasshopper

    M. strumosus Morse · swollen spurthroat grasshopper

    M. symmetricus Morse · symmetrical spurthroat grasshopper

    M. tepidus Morse · southern obovatewinged grasshopper

    M. tequestae Hubbell · Tequesta spurthroat grasshopper

Mermiria

    M. bivittata · twostriped mermiria grasshopper

    M. intertexta Scudder · eastern mermiria grasshopper

M. picta (F. Walker) · lively mermiria grasshopper

Metaleptea

    M. brevicornis (Johannson) · clippedwing grasshopper

Orphulella

    O. pelidna (Burmeister) · spottedwinged grasshopper

Pardalophora

    P. phoenicoptera (Burmeister) · orangewinged grasshopper

Paroxya

    P. atlantica Scudder · Atlantic grasshopper

    P. clavuliger (Serville) · olivegreen swamp grasshopper

Psinidia

    P. fenestralis (Serville) · longhorn bandwinged grasshopper

Romalea

    R. microptera (Beauvois) · eastern lubber grasshopper

Schistocerca

    S. alutacea (Harris) · rusty grasshopper

    S. americana (Drury) · American grasshopper

    S. ceratiola Hubbell and Walker · rosemary grasshopper

    S. damnifica (Saussure) · mischievous grasshopper

    S. obscura (Fabricius) · obscure birdwing grasshopper

Spharagemon

    S. bolli Scudder · Boll's grasshopper

    S. crepitans (Saussure) · crepitating grasshopper

    S. cristatum (Scudder) · ridgeback sand grasshopper

    S. marmorata (Scudder) · marbled grasshopper

Stenacris

    S. vitreipennis (Marschall) · glassywinged toothpick grasshopper

Syrbula

    S. admirabilis (Uhler) · handsome grasshopper

Trimerotropis

    T. maritima (Harris) · seaside grasshopper

Glossary

abdomen. The third or hindmost division of the insect body, the other divisions being the **head** and **thorax.**

annual. Plant that lives for only one growing season.

antenna (pl. antennae). Paired, segmented, elongate sensory structures located on the head.

anterior. Referring to the front or forward position.

apical. Referring to the apex, or tip.

appendage. A projecting part of the animal body, such as a limb or antenna; uaually paired.

biennial. Plant that lives for two growing seasons, with vegetative growth in the first season and reproduction in the second season.

biodiversity. The diversity of life forms in an ecosystem, usually based on the number of species, or species richness.

biomass. Amount of living matter.

broadleaf. A plant with a flat, wide leaf structure.

canopy. The uppermost leafy layer of a forest; overstory.

carina (pl. carinae). An elevated, longitudinal ridge on the **pronotum.**

cercus (pl. cerci). Paired appendages near the hind end of the abdomen.

crepitation. Crackling sound produced during flight by rubbing the undersurface of the forewings against the veins of the hind wings.

crenulate. Wavy; scalloped.

distal. Situated away from the point of attachment, as opposed to **proximal.**

dorsal. Referring to the upper surface, which in grasshoppers is the "back."

endemic. Sometimes used to indicate species that are native to an area and found nowhere else; **precinctive.**

exotic. Not naturally found in an area; alien.

femur (pl. femora). The third and stoutest segment of the leg; the "thigh."

forb. An **herb** other than a **grass**; broadleaf annual or short-lived plant.

forewing. The front pair of wings, closest to the head, and in grasshoppers usually covering the hind wings when the insect is not in flight.

frontal costa. A broad, flat ridge on the face of the grasshopper, also called **frontal ridge.**

frontal ridge. A broad, flat ridge or elevated region on the front of the head, extending from the eyes to above the mouth.

furcula. A forked process or projection at the rear of male grasshoppers that overlays the **supra-anal plate;** only the projections are visible, so it appears to be paired structures.

genus (pl. genera). A taxonomic group that includes closely related **species** and is itself included within subfamilies, which are within families. The scientific name of a species is the genus plus the species name; for example, *Melanoplus* is a genus and *Melanoplus nanciae* is a species within that genus.

grass. Common name for plants in the family Graminae; monocotyledonous plants (i.e., one cotyledon or leaf emerging from the seed) with narrow leaves.

gregarious. Living and moving as a group, without the division of labor of true social insects.

hammock. A forested area that is dominated by **broadleaf** trees.

head. The forward, or **anterior,** division of the body, followed by the **thorax** and **abdomen.**

herb. Nonwoody, short-lived plant.

herbivore. A plant-feeding animal.

hind wings. The second pair or **posterior** pair of wings; in grasshoppers capable of flight, these are the larger wings, but they fold and are hidden beneath the forewings when the grasshopper is at rest.

horizontal. Oriented in a plane parallel to the horizon, or along the length of the body, as opposed to **vertical.**

instar. The stage of the grasshopper between molts; this term is applied to grasshopper **nymphs,** usually in combination with a number to indicate whether it is early in development (e.g., 1st instar) or late (e.g., 5th instar).

lateral. Relating to the side.

leg. Appendage associated with the **thorax** and used for **terrestrial** locomotion.

mesic. Relating to a moist but not wet environment.

metamorphosis. Change in body form (e.g., change from a caterpillar to a moth).

molt. To cast off the outgrown exterior body covering, a process that occurs between **instars.**

morphology. Form or structure of an organism; appearance.

nutrient cycling. The movement of nutrients, usually mineral in nature, among **trophic levels.**

nymph. The young of grasshoppers; immature insects that resemble the adult in body form, differing mainly in wing development and ability to reproduce.

pallium. An erect conical structure at the tip of the **supra-anal plate.**

posterior. Referring to the back or rear position.

precinctive. Native or **endemic** species, not known to occur elsewhere.

pronotum. The upper, or **dorsal,** section of the **prothorax,** which is the first of the three segments of the **thorax;** in grasshoppers the pronotum largely hides the mesonotum and metanotum, the upper portions of the middle and third segments of the **thorax.**

prosternal spine. A small spur or spine located on the undersurface of the **prothorax;** a spine protruding from between the front legs, especially in the subfamily Cyrtacanthacridinae.

prothorax. The first, or **anterior,** segment of the **thorax.**

proximal. Located closest to the point of attachment or to the center of the body; opposite of **distal.**

shrub. A woody plant, typically with several major stems.

species. Groups of actually or potentially interbreeding populations that are isolated from other groups, usually by behavior or geography, and thus unable to mate with those other groups. See also **genus.**

speciation. The evolution of species.

stridulation. Creaking sound that grasshoppers produce while not in flight made by rubbing the inner surface of the hind **femur** on the edges of the **forewing.**

spine. Elongate, pointed structure; in the subfamily Cyrtacanthacridinae, the "spurthroated" grasshoppers, a large spine or spur called the **pro-**

sternal spine is present between, or slightly in front of, the front legs, but it is sometimes reduced to a bluntly rounded elevation; small but sharply pointed spines also occur in rows along the **tibiae.**

spur. See **spine**, above.

stridulate. To make a noise by rubbing two surfaces or structures together.

subgenital plate. A plate at the tip of the **abdomen** that covers the genital area from below; it tends to be curved upward and scoop-shaped.

supra-anal plate. A plate at the tip of the **abdomen** that covers the genital area from above; it is flat and triangular or shield-shaped.

sword-shaped antenna. Antenna with flattened segments, widening from the base and then narrowing toward the tip.

tarsus (pl. tarsi). The **distal** segment of the leg, the "foot."

tegmen (pl. tegmina). The thickened **forewings** of grasshoppers and also of crickets, cockroaches, and praying mantids.

terrestrial. Of or relating to land, as opposed to water.

thorax. The second, or middle, of the three major body divisions.

tibia (pl. tibiae). The long, thin fourth segment of the leg, between the **femur** and **tarsus.**

transverse. Lying across, crosswise, at right angles to.

tree. A woody plant, typically with one major stem.

trophic level. Position in a food chain or food web.

tympanum (pl. tympana). A tightly stretched membrane covering the "hearing" or auditory organ, and located on the side of the **abdomen.**

ovipositor. The structures located at the tip of the **abdomen** in females and used to deposit eggs.

perennial. A plant living several years, not dying soon after reproduction.

precinctive. Unique to an area, indigenous.

pronotum. The covering of the upper surface of the **prothorax.**

prothorax. The first segment of the **thorax.**

terminal. Of or relating to an end.

understory. Trees and shrubs below the canopy of a forest.

wing pad. The partly developed wings located at the juncture of the **thorax** and **abdomen.**

vertical. Oriented up and down, as opposed to **horizontal.**

ventral. Refers to the underside, or below.

xeric. Relating to a very dry environment.

References

Blatchley, W. S. 1920. *Orthoptera of Northeastern North America.* Indianapolis: Nature Publishing Company.

Capinera, J. L., and T. S. Sechrist. 1982. Grasshoppers (Acrididae) of Colorado: identification, biology and management. *Colorado Agricultural Experiment Station Bulletin 584S.*

Chapman, R. F., and A. Joern, eds. 1990. *Biology of Grasshoppers.* New York: John Wiley.

Dakin, M. E., and K. L. Hays. 1970. A synopsis of Orthoptera (*sensu lato*) of Alabama. *Auburn University Agricultural Experiment Station Bulletin 404.*

Deyrup, M. 1996. Two new grasshoppers from relict uplands of Florida (Orthoptera: Acrididae). *Transactions of the American Entomological Society* 122:199–211.

Friauf, J. J. 1953. An ecological study of the Dermaptera and Orthoptera of the Welaka area in northern Florida. *Ecological Monographs* 23:79–126.

Helfer, J. R. 1972. *How to Know the Grasshoppers, Cockroaches, and Their Allies,* 2nd ed. Dubuque: W. C. Brown Company.

Hubbell, T. H. 1932. A revision of the Puer group of the North American genus *Melanoplus,* with remarks on the taxonomic value of the concealed male genitalia in the Cyrtacanthacridinae (Orthoptera, Acrididae). *University of Michigan Museum of Zoology Miscellaneous Publication 23.*

Myers, R. L., and J. J. Ewel, eds. 1990. *Ecosystems of Florida.* Orlando: University of Central Florida Press.

Otte, D. 1981. *Acrididae: Gomphocerinae and Acridinae.* Vol. 1, *The North American Grasshoppers.* Cambridge: Harvard University Press.

Otte, D. 1984. *Acrididae: Oedipodinae.* Vol. 2, *The North American Grasshoppers.* Cambridge: Harvard University Press.

Richman, D. B., D. C. Lightfoot, C. A. Sutherland, and D. J. Ferguson. 1993. A manual of the grasshoppers of New Mexico. Orthoptera: Acrididae and Romaleidae. *New Mexico State University Cooperative Extension Service Handbook No. 7.*

Index

Entries are alphabetized according to the scientific and common names of the species; for example, *Archurum carinatum* is listed under both *Archurum carinatum* and longheaded toothpick grasshopper. Numbers in boldface indicate pages on which illustrations—line drawings, photographs, and distribution maps—accompany text for a given entry.

About the Authors

John L. Capinera is professor and chairman of the Department of Entomology and Nematology at the University of Florida. Although he has broad interests in insect ecology and in crop pest management, for more than twenty years he has been particularly fascinated by grasshoppers and has published numerous articles on grasshopper biology.

Clay W. Scherer and Jason M. Squitier, former graduate students in entomology at the University of Florida, conducted research on the relationship of grasshopper diversity and abundance to plant availability and community structure.